The HVAC/R Professional's Field Guide to

# Universal R-410A
# Safety & Training

## Delta-T Solutions

John Tomczyk              Joe Nott              Dick Shaw

Published by:

**ESCO press**

Mount Prospect, Illinois

**ISBN 1–930044–12–7**

This book was written as a general guide. The authors and publishers have neither liability nor can
they be responsible to any person or entity for any misunderstanding, misuse, or misapplication that
would cause loss or damage of any kind, including loss of rights, material, or personal injury alleged
to be caused directly or indirectly by the information contained in this book.

Cover photo supplied by **Honeywell International**

This image may not be reproduced without written authorization of Honeywell. Gentron ® and AZ–20 ® are
registered trademarks of Honeywell. Honeywell has not reviewed the material contained herein, and assumes
no liability or responsibility for its use.

Printed in the United States of America
7654321

The AC&R Safety Coalition develops and establishes safe work practices and paradigms for Heating, Ventilation, Air Conditioning, and Refrigeration personnel. This is accomplished through education, training and certification programs that address the safety needs of our industry.

## Founding Members:

# Bibliographical Acknowledgments

**We wish to thank the following organizations for whose material used in research made this project possible.**

Air Conditioning Contractors of America (ACCA)
Air Conditioning and Refrigeration Institute (ARI)
Amana
The Air Conditioning, Heating and Refrigeration News
American Society of Heating, Refrigeration & Air Conditioning Engineers, Inc. (ASHRAE)
Blissfield Manufacturing Company
Bohn Heat Transfer Company
BVA Oils
Carlyle
Carrier
Castrol
Chevron
Copeland
Danfoss, Inc.
The Delfield Company
DuPont Chemicals
ESCO Institute (Educational Standards Corp)
Frigidaire Company
General Motors
Goodman Manufacturing
Honeywell
HVAC Excellence
Industrial Technology Excellence (ITE)
Johnson Controls
Lennox
Merit Mechanical Systems, Inc.
Mobil
National Refrigerants
Newsweek
Plumbing Heating and Cooling Contractors (PHCC)
Refrigeration Research, Inc.
Refrigeration Service Engineers Society (RSES)
Rheem
Rhuud
Ritchie Engineering
Robinair (SPX)
Scientific American
Tecumseh Corporation
Thermal Engineering Company
TIF Corporation (SPX)
Time
Trane
United States Department of Transportation
United States Environmental Protection Agency (EPA)
York

# Table of Contents

## Sections

## I R–410A and the R–22 Phase–Out

## II Refrigeration and Air Conditioning Systems Fundamentals

# III  Refrigerant Chemistry and Applications

# IV      Refrigeration Oils and Their Applications

# V      Safety

# VI    Appendix I

# VII   Appendix II

## Universal R–410A Safety

# Preface

This certification manual was written to assist in the training and certification of HVACR technicians for proper safety, handling and application of R–410A refrigerant. The manual was written by two current and one Emeriti faculty members of the HVACR Department of Ferris State University.

The program is written on the belief that the solution to transition to environmentally safer refrigerants and oils, while keeping the public and technicians out of harms way, is education and training. This value–added program contains practical applications of refrigeration and air conditioning system technology, fundamentals of refrigerants and oils, and the characteristics of R–410A, a refrigerant that deserves safety consideration.

This project was conducted in cooperation with numerous manufacturers and associations, most of which are listed with the acknowledgments to this manual. Their assistance made the solutions and safety portions of this manual possible. At the time of printing, the information on refrigerants and oils was the current technology.

# *R–410A and the R–22 Phase–Out*

## Background:

It is widely accepted that chlorine based refrigerants contribute to the depletion of the earth's stratospheric ozone. In recent years, the air conditioning and refrigeration industry has supported global efforts to transition to safer non–chlorine based refrigerants. In the developing countries of the world, CFC–12 (R–12) refrigerant, which was widely used since the 1930's, is today phased out and replaced with non–ozone depleting refrigerants. HCFCs, (including R–22) that have been widely used in air conditioning and refrigeration applications since the 1940's, are also being phased out. The technological changes that continue to evolve with refrigerants, compressor design, highly refined refrigeration oils and increased efficiency is truly revolutionary. The challenges confronting the refrigeration and air conditioning industry continue to unfold as we provide industrial cooling, comfort, food preservation and the "quality of life" needed for our society. This manual addresses one of these challenges; the transition from R–22 to R–410A.

Based on the 1974 Molina–Rowland theory that chlorine and bromine were responsible for depleting the earth's ozone layer that protects us from ultraviolet radiation, numerous global actions have taken place to reverse this environmental problem. Let's look at some of these significant actions:

- 1978 U.S. bans all non-essential aerosols containing chlorine or bromine.

- 1978 global warming concerns come into view.

- 1987 the U.S. and 22 other countries sign the original Montreal Protocol establishing timetables and phase-out schedules for CFCs and HCFCs.

- 1990 The Clean Air Act (CAA) signed in the U.S. calling for refrigerant, production reductions, recycling and emission reduction and the eventual phase-out of CFCs and HCFCs.

- 1992 unlawful to vent CFCs and HCFCs into the atmosphere.

- 1994 technician certification required for purchasing and handling of CFCs and HCFCs.

- 1995 unlawful to vent alternate (substitute) refrigerants such as HFCs, into the atmosphere.

- 1996 phase-out of CFC refrigerant production in the U.S.

- 1996 cap HCFC production levels.

- 1997 Kyoto Protocol established in response to global warming concerns.

- 2010 phase-out HCFC-22 (R-22) for new equipment.

- 2020 Phase-out HCFC-22 production.

## HCFC Phaseout Schedule

The following italicized statements are condensed
and reprinted from the U. S. EPA web site.

*All developed countries that are Parties to the Montreal Protocol are subject to a cap on their consumption of hydrochlorofluorocarbons (HCFCs).*

*Consumption is calculated by the following formula:*
*consumption = production plus imports minus exports.*

*The cap is set at 2.8% of that country's 1989 chlorofluorocarbon consumption + 100% of that country's 1989 HCFC consumption. (Quantities of chemicals measured under the cap are ODP–weighted, which means that each chemical's relative contribution to ozone depletion is taken into account.)*

*Under the Montreal Protocol, the U.S. and other developed nations are obligated to achieve a certain percentage of progress towards the total phaseout of HCFCs, by certain dates. These nations use the cap as a baseline to measure their progress towards achieving these percentage goals.*

*The following table shows the U.S. schedule for phasing out its use of HCFCs in accordance with the terms of the Protocol. The Agency intends to meet the limits set under the Protocol by accelerating the phaseout of HCFC–141b, HCFC–142b and HCFC–22. These are the most damaging of the HCFCs. By eliminating these chemicals by the specified dates, the Agency believes that it will meet the requirements set by the Parties to the Protocol. The third and fourth columns of the table show how the United States will meet the international obligations described in the first two columns.*

Since the production levels are based on caps, rising production levels of HCFCs has triggered an accelerated phase–out of some HCFCs by manufacturers of new air conditioning equipment, prior to the established phase–out schedule.

**See Phase–out chart**

## PHASE–OUT CHART

| Montreal Protocol | | United States | |
| --- | --- | --- | --- |
| Year by which Developed Countries Must Achieve % Reduction in Consumption | % Reduction in Consumption, Using the Cap as a Baseline | Year to be Implemented | Implementation of HCFC Phaseout through Clean Air Act Regulations |
| 2004 | 35.0% | 2003 | No production and no importing of HCFC–141b |
| 2010 | 65% | 2010 | No production and no importing of HCFC–142b and HCFC–22, except for use in equipment manufactured before 1/1/2010 (so no production or importing for NEW equipment that uses these refrigerants) |
| 2015 | 90% | 2015 | No production and no importing of any HCFCs, except for use as refrigerants in equipment manufactured before 1/1/2020 |
| 2020 | 99.5% | 2020 | No production and no importing of HCFC–142b and HCFC–22 |
| 2030 | 100% | 2030 | No production and no importing of any HCFCs |

Due to environmental and competitive pressure, HCFCs are being phased–out. In response, many manufacturers are building air conditioning equipment using HFC based R–410A. It is important that contractors and technicians understand the safety, safe handling, proper charging, operating characteristics and proper applications of this refrigerant blend.

As we approach the next stage and comply with these global and national provisions and regulations calling for the elimination of all ozone depleting substances, we need to prepare ourselves.

## Regulation and Change:

Public pressures that resulted in the Montreal Protocol and regulations imposed by the CAA have resulted in our industry's transition to safer refrigerants. Numerous other factors such as global warming, energy utilization, developments in compressor design and refrigeration oils also continue to create change.

Global warming is a challenge that may see increased attention as our industry phases into newer refrigerants and more efficient equipment. The Kyoto Protocol that was established in 1997 calls for the reduction of greenhouse gases by an average of 5.23% from 1990 levels in developing countries. While only a few nations have ratified the Kyoto Protocol, many countries are reacting strongly and our industry may be challenged to look to alternate refrigerants that reduce global warming. Measurements of global warming such as the Total Equivalent Warming Impact (TEWI) take into consideration the direct and indirect effects of global warming, and can play an increased part in the selection of new refrigerants and system performance.

The development of the scroll compressor and the rapid adoption from the reciprocating compressor has opened the door to new refrigerants and made our industry's challenge easier. The scroll compressor not only is more efficient; it also is able to accommodate considerably higher pressures that are inherent in R–410A.

In 2006 the U.S. Department of Energy is scheduled to require that air conditioner efficiencies be raised from 10 SEER (Seasonal Energy Efficiency Ratio) to 12 SEER or higher. ASHRAE 90.1 standard is calling for increased efficiency level in commercial equipment to be increased by as much as 20%.

The direct and indirect impact of R–410A on global warming must be considered. The direct impact of R–410A is that it has a slightly higher global warming potential (GWP) than R–22. The indirect impact of R–410A is that because of its increased efficiency, R–410A systems use less energy, thereby reducing carbon dioxide emissions from power plants. The TEWI should be lower. There will likely be increased pressure on our industry to not only transition to safer refrigerants, but to further reduce refrigerant emissions, produce higher efficient equipment and maintain these systems at their optimum level of efficiency. That is our challenge.

The Federal Clean Air Act calls for the phase out of HCFC based R–22 with no production or importing beginning in 2020. However, manufacturers of air conditioning equipment must discontinue the use of HCFC based R–22 in new equipment by January 1, 2010. Many manufacturers have already begun the phase out process and are building specially designed systems that utilize the HFC based R–410A refrigerant. The newly designed R–410A systems employ thicker walled tubing, newly developed compressors, components and a higher grade oil that require different installation and service procedures.

## The Future:

HFCs such as R–410A, R–407C and R–134a are the refrigerants of choice for this generation. These refrigerants solve the initial problem of stratospheric ozone depletion, but they still have some global warming potential. It is important that we recognize this as an evolutionary process. As we continue to transition to R–410A, R–407C and R–134a, we could see technological changes and pressures that may bring newer refrigerants and more transition. Change is inevitable. This is not the last chapter.

Regulations of the CAA prohibit venting of HFC refrigerants, and we can expect increased emphasis in the areas of refrigerant containment and recycling of all refrigerants. With increased attention to global warming and climate change, we may see a new family of refrigerants and changes in refrigeration and air conditioning systems. Energy shortages, along with higher utility bills may bring increased demand for maintenance and service procedures that guarantee HVACR systems operate at their peak performance.

## Safety and R–410A:

R–410A is a binary (two part) near–azeotropic mixture and is presently marketed under the brand names of AZ–20 "Puron" or "Suva.." Chapter 3, "Refrigerant Chemistry and Applications," of this manual provides a good foundation and explanation of the properties of R–410A.

These newly manufactured R–410A air conditioning systems will require contractors and technicians to shift to different tools, equipment and safety standards when installing or changing out (retrofitting) older split A/C systems and repairing systems in the field. Chapter 5, "Safety," will provide the background, foundation and procedures for making this shift.

Mr. Slim. Com
410 additional info

We need to know that R–410A can only be used in equipment specifically designed and constructed for R–410A. R–22 systems cannot be retrofitted, without major component upgrades, to R–410A. We need to know that R–410A operates at considerably higher pressures and why special cylinders, gauges and recovery equipment are necessary. And most importantly, we need to know how to safely handle R–410A. Since R–410A utilizes Polyol Ester (POE) based oils and not mineral oil, we need to know how to properly install and service these systems that are not as forgiving relative to moisture.

R–410A operates at significantly higher pressures and refrigeration capacity. It is important to know why all refrigerant flow controls, feed devices, driers and compressors are specifically designed for R–410A.

The following sections will prepare you for successful completion of the "AC&R Safety Coalition R–410A Certification". This certification will show evidence of your professional ability to safely handle and work with this new generation of refrigerants.

# Refrigeration and A/C Systems Fundamentals

## Objectives

- ◆ Describe condensing and evaporating pressure.
- ◆ Explain the liquid and vapor states of refrigerants.
- ◆ Describe a superheated vapor and a subcooled liquid.
- ◆ List the components of the basic vapor compression system.
- ◆ Demonstrate the formula for calculating compression ratio.
- ◆ Select the proper components to be used with the higher pressure refrigerant R-410A.

## Vapor Compression Refrigeration System

Refrigeration is defined as that branch of science which deals with the process of reducing and maintaining the temperature of a space or materials below the temperature of the surroundings. To accomplish this, heat must be removed from the refrigerated body and transferred to another body.

The typical vapor compression refrigeration system shown in **figure 2–1** can be divided into two pressures: condensing (high side) and evaporating (low side). These pressures are divided or separated in the system by the compressor discharge valve and the metering device. Listed in the chart below are field service terms often used to describe these pressures.

## Condensing Pressure

The condensing pressure is the pressure at which the refrigerant changes state from a vapor to a liquid. This phase change is referred to as *condensation.* This pressure can be read directly from a pressure gauge connected anywhere between the compressor discharge valve and the entrance to the metering device, assuming there is negligible pressure drop. In reality, line and valve friction and the weight of the liquid itself cause pressure drops from the compressor discharge to the metering device. If the true condensing pressure is needed, the technician must measure the pressure as close to the condenser as possible to avoid these pressure drops. This pressure is usually measured on smaller systems near the compressor valves. On small systems, it is not critical where a technician places the pressure gauge (as long as it is on the high side of the system), because pressure drops are negligible. The pressure gauge reads the same no matter where it is on the high side of the system if line and valve losses are negligible.

| Condensing Pressure | Evaporating Pressure |
| --- | --- |
| High side pressure | Low side pressure |
| Head pressure | Suction pressure |
| Discharge pressure | Back pressure |

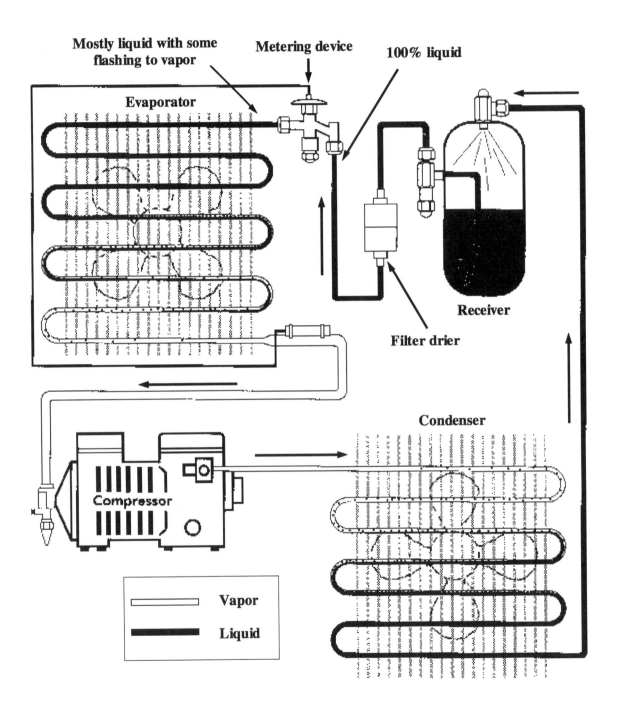

**Figure 2-1 Typical compression refrigeration system**

## Evaporating Pressure

The evaporating pressure is the pressure at which the refrigerant changes state from a liquid to a vapor. This phase change is referred to as evaporation or vaporizing. A pressure gauge placed anywhere between the metering device outlet and the compressor (including compressor crankcase) will read the evaporating pressure. Again, negligible pressure drops are assumed. In reality, there will be line and valve pressure drops as the refrigerant travels through the evaporator and suction line.

The technician must measure the pressure as close to the evaporator as possible to get a true evaporating pressure. On small systems where pressure drops are negligible, this pressure is usually measured near the compressor. Gauge placement on small systems is usually not critical as long as it is placed on the low side of the refrigeration system, because the refrigerant vapor pressure acts equally in all directions. If line and valve pressure drops become substantial, gauge placement becomes critical. In larger more sophisticated systems, gauge placement is more critical because of associated line and valve pressure losses. If the system has significant line and valve pressure losses, the technician must place the gauge as close as possible to the component that requires a pressure reading.

## Refrigerant States and Conditions

Modern refrigerants exist either in the vapor or liquid state. Refrigerants have such low freezing points that they are rarely in the frozen or solid state. Refrigerants can co–exist as vapor and liquid as long as conditions are right. Both the evaporator and condenser house liquid and vapor refrigerant simultaneously if the system is operating properly. Refrigerant liquid and vapor can exist in both the high or low pressure sides of the refrigeration system.

Along with refrigerant pressures and states are refrigerant conditions. Refrigerant conditions can be ***saturated, superheated,*** or ***subcooled.***

## Saturation

Saturation is usually defined as a temperature. The saturation temperature is the temperature at which a fluid changes from liquid to vapor or vapor to liquid. At saturation temperature, liquid and vapor are called saturated liquid and saturated vapor, respectively. Saturation occurs in both the evaporator and condenser. At saturation, the liquid experiences its maximum temperature for the pressure, and the vapor experiences its minimum temperature. However, both liquid and vapor are at the same temperature for the given pressure when saturation occurs. Saturation temperatures vary with different refrigerants and pressures. All refrigerants have different vapor pressures. It is vapor pressure that is measured with a gauge.

## Vapor Pressure

Vapor pressure is the pressure exerted on a saturated liquid. Any time saturated liquid and vapor are together (as in the condenser and evaporator), vapor pressure is generated. Vapor pressure acts equally in all directions and affects the entire low or high side of a refrigeration system.

As pressure increases, saturation temperature increases; as pressure decreases, saturation temperature decreases. Only at saturation are there pressure/temperature relationships for refrigerants. **Table 2–1** shows the pressure/temperature relationship at saturation for refrigerant R–134a. If one attempts to raise the temperature of a saturated liquid above its saturation temperature, vaporization of the liquid will occur. If one attempts to lower the temperature of a saturated vapor below its saturation temperature, condensation will occur. Both vaporization and condensation occur in the evaporator and condenser, respectively.

The heat energy that causes a liquid refrigerant to change to a vapor at a constant saturation temperature for a given pressure is referred to as *latent heat*. Latent heat is the heat energy that causes a substance to change state without changing the temperature of the substance. Vaporization and condensation are examples of a latent heat process.

## Superheat

Superheat always refers to a vapor. A superheated vapor is any vapor that is above its saturation temperature for a given pressure. In order for vapor to be superheated, it must have reached its 100% saturated vapor point. In other words, all of the liquid must be vaporized for superheating to occur; the vapor must be removed from contact with the vaporizing liquid. Once all the liquid has been vaporized at its saturation temperature, any addition of heat causes the 100% saturated vapor to start superheating. This addition of heat causes the vapor to increase in temperature and gain sensible heat. Sensible heat is the heat energy that causes a change in the temperature of a substance. The heat energy that superheats vapor and increases its temperature is sensible heat energy. Superheating is a sensible heat process. Superheated vapor occurs in the evaporator, suction line, and compressor.

| Temperature (°F) | Pressure (psig) | Temperature (°F) | Pressure (psig) |
|---|---|---|---|
| −10 | 1.8 | 25 | 21.7 |
| −9 | 2.2 | 26 | 22.4 |
| −8 | 2.6 | 27 | 23.2 |
| −7 | 3.0 | 28 | 24.0 |
| −6 | 3.5 | 29 | 24.8 |
| −5 | 3.9 | 30 | 25.6 |
| −4 | 4.4 | 31 | 26.4 |
| −3 | 4.8 | 32 | 27.3 |
| −2 | 5.3 | 33 | 28.1 |
| −1 | 5.8 | 34 | 29.0 |
| 0 | 6.2 | 35 | 29.9 |
| 1 | 6.7 | 40 | 34.5 |
| 2 | 7.2 | 45 | 39.5 |
| 3 | 7.8 | 50 | 44.9 |
| 4 | 8.3 | 55 | 50.7 |
| 5 | 8.8 | 60 | 56.9 |
| 6 | 9.3 | 65 | 63.5 |
| 7 | 9.9 | 70 | 70.7 |
| 8 | 10.5 | 75 | 78.3 |
| 9 | 11.0 | 80 | 86.4 |
| 10 | 11.6 | 85 | 95.0 |
| 11 | 12.2 | 90 | 104.2 |
| 12 | 12.8 | 95 | 113.9 |
| 13 | 13.4 | 100 | 124.3 |
| 14 | 14.0 | 105 | 135.2 |
| 15 | 14.7 | 110 | 146.8 |
| 16 | 15.3 | 115 | 159.0 |
| 17 | 16.0 | 120 | 171.9 |
| 18 | 16.7 | 125 | 185.5 |
| 19 | 17.3 | 130 | 199.8 |
| 20 | 18.0 | 135 | 214.8 |
| 21 | 18.7 | | |
| 22 | 19.4 | | |
| 23 | 20.2 | | |
| 24 | 20.9 | | |

*Table 2–1*
*R–134a saturated vapor/liquid pressure/ temperature chart*

## Subcooling

Subcooling always refers to a liquid at a temperature below its saturation temperature for a given pressure. Once all of the vapor changes state to 100% saturated liquid, further removal of heat will cause the 100% liquid to drop in temperature or lose sensible heat. Subcooled liquid results. Subcooling can occur in both the condenser and liquid line and is a sensible heat process.

A thorough understanding of pressures, states, and conditions of the basic refrigeration system enables the service technician to be a good systematic troubleshooter. It is not until then that a service technician should even attempt systematic troubleshooting.

## Basic Refrigeration System Components

## Compressor

One of the main functions of the compressor is to circulate refrigerant. Without the compressor as a refrigerant pump, refrigerant could not reach other system components to perform its heat transfer functions. (**Fig. 2–2**) The compressor also separates the high pressure from the low pressure side of the refrigeration system. A difference in pressure is mandatory for fluid (gas or liquid) flow, and there could be no refrigerant flow without this pressure separation.

Another function of the compressor is to elevate or raise the temperature of the refrigerant vapor above the ambient (surrounding) temperature. This is accomplished by adding work, or heat of compression, to the refrigerant vapor during the compression cycle. The pressure of the refrigerant is raised, as well as its temperature. By elevating the refrigerant temperature above the ambient temperature, heat absorbed in the evaporator and suction line, and any heat of compression generated in the compression stroke can be rejected to this lower temperature ambient. Most of the heat is rejected in the discharge line and the condenser. Remember, heat flows from hot to cold, and there must be a temperature difference for any heat transfer to take place. The temperature rise of the refrigerant during the compression stroke is a measure of the increased internal kinetic energy added by the compressor.

The compressor also compresses the refrigerant vapors, which increases vapor density. This increase in density helps pack the refrigerant gas molecules together, which helps in the condensation or liquefication of the refrigerant gas molecules in the condenser once the right amount of heat is rejected to the ambient. The compression of the vapors during the compression stroke is actually preparing the vapors for condensation or liquefication.

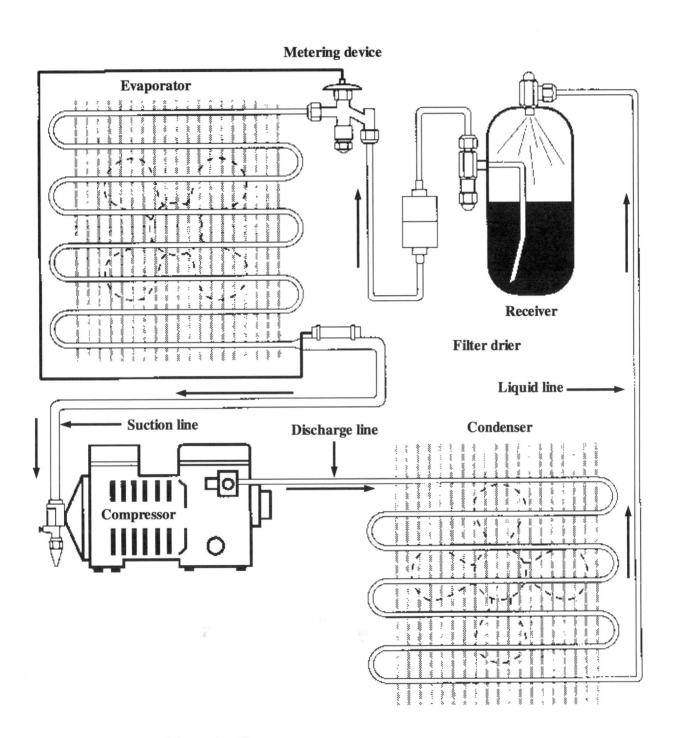

**Fig. 2-2 Basic Refrigeration System**

## Compression Ratio

Compression ratio is the term used with compressors to describe the difference between the low and high sides of the compression cycle. By knowing the compression ratio, a technician can tell how efficiently a system is operating. When the compression ratio is low, lower power consumption and higher efficiencies will be experienced. Compression ratio is calculated as follows in **Equation #1**:

### Equation #1
*Compression ratio = $\dfrac{\text{Absolute discharge pressure}}{\text{Absolute suction pressure}}$*

The reason that absolute pressures rather than gauge pressures are used in the compression ratio equation is to keep the compression ratio a positive and meaningful number (e.g. if the suction pressure fell into a vacuum). Remember, absolute pressures are simply gauge pressures plus atmospheric pressure (**Equation #2**). A compression ratio of 7 to 1 (7:1) simply implies that the discharge pressure is 7 times greater than the suction pressure. As one can see, an increase in discharge pressure or a decrease in suction pressure will make the compression ratio higher, thus the system less efficient. On the other hand, a decrease in discharge pressure or an increase in suction pressure will make the compression ratio lower, thus the system more efficient and draw less power. Both high and low side operating pressures are approximately 50% to 70% higher for R–410A systems when compared to R–22 systems. This results in similar compression ratios and power consumption for the two refrigerants.

### Equation #2
**Absolute pressure = (Gauge pressure + Atmospheric pressure)**

## R–410A Considerations

Manufacturers have redesigned their compressors with increased wall thickness due to the higher pressures associated with R–410A. **(A compressor designed for R–22 should never be used with R-410A** Also compressor internal pressure relief (IPR) settings are different for R–22 and R–410A systems. The IPR will open at a pressure of 375–450 psig for R–22 systems and a pressure of 550–625 psig for R–410A systems. Although suction and discharge pressures are 50–70% greater with R–410A than R–22 the discharge temperature of R–410A is lower due to its higher vapor heat capacity.

Systems that use R–410A will require a change in the high and low pressure switch settings due to the increased pressure of the refrigerant. (Automatic reset controls are now being used). The high pressure switch will now open at 610 psig plus or minus 10 psig and close at 500 psig plus or minus 15 psig. The low pressure control will open at 50 psig.

## Discharge Line

One function of the discharge line is to carry the high pressure superheated vapor from the compressor discharge valve to the entrance of the condenser. The discharge line also acts as a desuperheater, cooling the superheated vapors that the compressor has compressed and giving that heat up to the ambient (surroundings). These compressed vapors contain all of the heat that the evaporator and suction line have absorbed, along with the heat of compression of the compression stroke. Any generated motor winding heat may also be contained in the discharge line refrigerant, which is why the beginning of the discharge line is the hottest part of the refrigeration system. On hot days when the system is under a high load and may have a dirty condenser, the discharge line can reach over 400° F. By de–superheating the refrigerant, the vapors will be cooled to the saturation temperature of the condenser. Once the vapors reach the condensing saturation temperature for that pressure, condensation of vapor to liquid will take place as more heat is lost.

## Condenser

The first passes of the condenser desuperheat the discharge line gases. This prepares the high pressure superheated vapors coming from the discharge line for condensation, or the phase change from gas to liquid. Remember, these superheated gases must lose all of their superheat before reaching the condensing temperature for a certain condensing pressure. Once the initial passes of the condenser have rejected enough superheat and the condensing temperature or saturation temperature has been reached, these gasses are referred to as 100% saturated vapor. The refrigerant is then said to have reached the 100% saturated vapor point, point #2, **figure 2–3.**

One of the main functions of the condenser is to condense the refrigerant vapor to liquid. Condensing is system dependent and usually takes place in the lower two–thirds of the condenser. Once the saturation or condensing temperature is reached in the condenser and the refrigerant gas has reached 100% saturated vapor, condensation can take place if more heat is removed. As more heat is taken away from the 100% saturated vapor, it will force the vapor to become a liquid or to condense. When condensing, the vapor will gradually phase change to liquid until 100% liquid is all that remains. This phase change, or change of state, is an example of a latent heat rejection process, as the heat removed is latent heat not sensible heat. The phase change will happen at one temperature even though heat is being removed. Note: An exception to this is a near–azeotropic blend of refrigerants where there is a temperature glide or range of temperatures when phase changing (see Chapter 3 on blend temperature glide). This one temperature is the saturation temperature corresponding to the saturation pressure in the condenser.

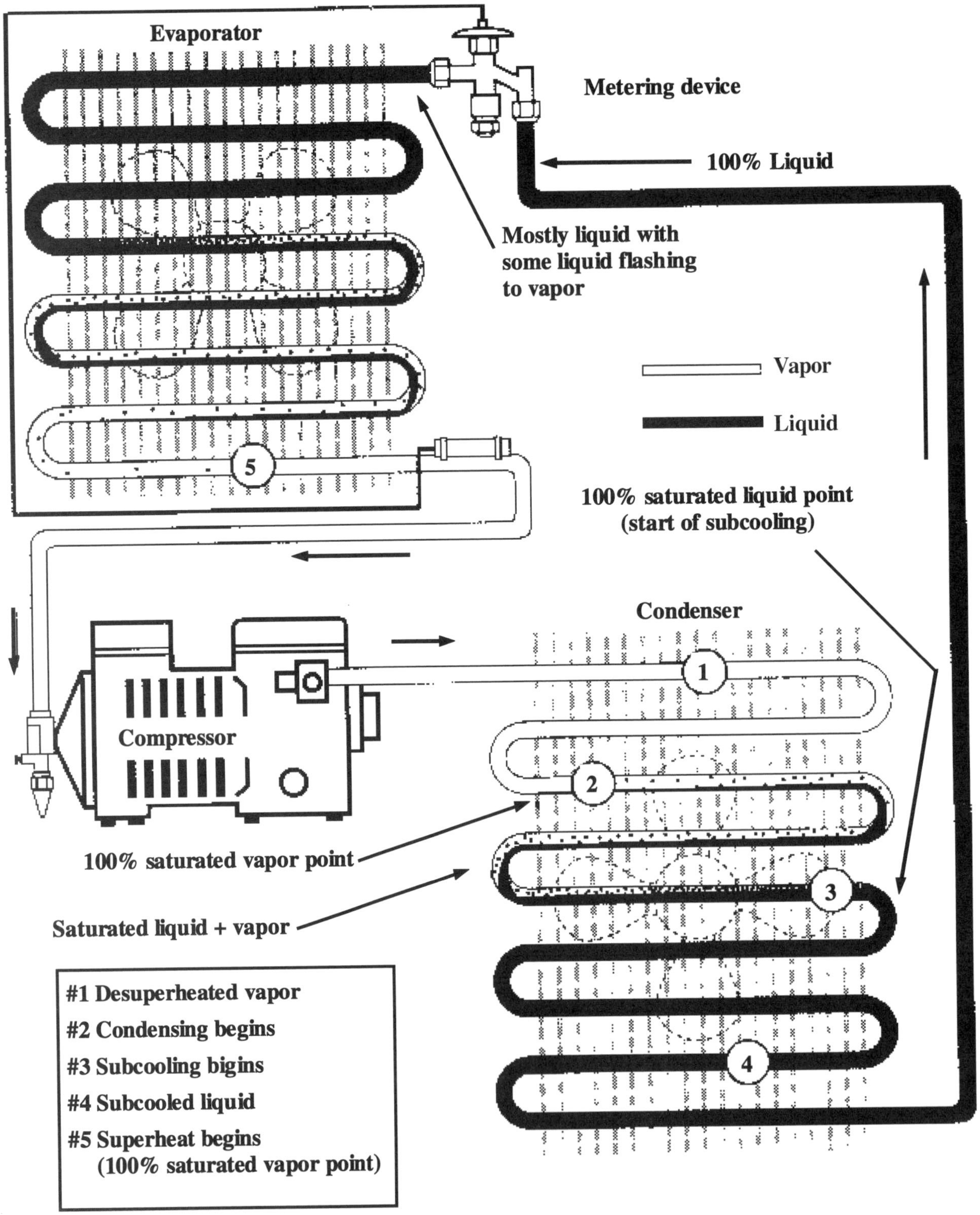

Evaporator

Metering device

100% Liquid

Mostly liquid with
some liquid flashing
to vapor

Vapor

Liquid

100% saturated liquid point
(start of subcooling)

Condenser

Compressor

100% saturated vapor point

Saturated liquid + vapor

#1 Desuperheated vapor
#2 Condensing begins
#3 Subcooling bigins
#4 Subcooled liquid
#5 Superheat begins
   (100% saturated vapor point)

**Fig. 2-3. Basic refrigeration system showing 100% saturated vapor and liquid points**

The last function of the condenser is to subcool the liquid refrigerant. Subcooling is defined as any sensible heat taken away from 100% saturated liquid. Technically, subcooling is defined as the difference between the measured liquid temperature and the liquid saturation temperature at a given pressure. Once the saturated vapor in the condenser has phase changed to saturated liquid, the 100% saturated liquid point has been reached. If any more heat is removed, the liquid will go through a sensible heat rejection process and lose temperature as it loses heat. The liquid that is cooler than the saturated liquid in the condenser is subcooled liquid. Subcooling is an important process, because it starts to lower the liquid temperature to the evaporator temperature. This will reduce flash loss in the evaporator so more of the vaporization of the liquid in the evaporator can be used for useful cooling of the product load.

## *R–410A Considerations*

Equipment designed for R–22 cannot withstand the higher pressure of R–410A. The condensing unit must be replaced with a specific model designed for R–410A.

## Receiver

The receiver acts as a surge tank. Once the subcooled liquid exits the condenser, the receiver stores the liquid. The liquid level in the receiver varies depending on whether the metering device is throttling opened or closed. Receivers are usually used on systems in which a thermostatic expansion valve (TXV or TEV) is used as the metering device. The subcooled liquid in the receiver may lose or gain subcooling depending on the surrounding temperature of the receiver. If the subcooled liquid is warmer than receiver surroundings, the liquid will reject heat to the surroundings and subcool even more. If the subcooled liquid is cooler than the receiver surroundings, heat will be gained by the liquid and subcooling will be lost.

A receiver bypass is often used to bypass liquid around the receiver and route it directly to the liquid line and filter drier. This bypass prevents subcooled liquid from sitting in the receiver and losing its subcooling. A thermostat with a sensing bulb on the condenser outlet controls the bypass solenoid valve by sensing liquid temperature coming to the receiver, **figure 2–4**. If the liquid is subcooled to a predetermined temperature, it will bypass the receiver and go to the filter drier.

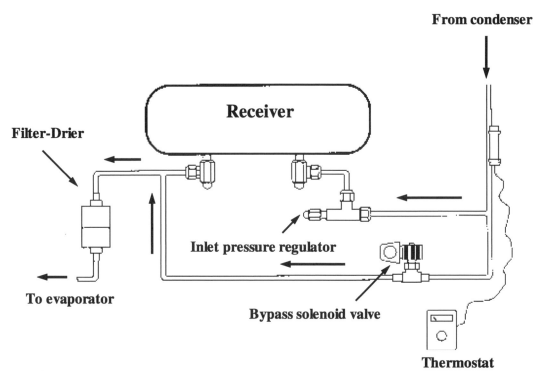

**Fig. 2-4. Receiver with thermostatically controlled liquid bypass.**

### Filter/Driers

A very small amount of moisture may be present in a refrigeration system, regardless of the care taken when evacuating and charging. If moisture is allowed to remain in a system, acid formation, corrosion, oil breakdown, sludge, or carbon build–up may result. Any of these contaminants may cause compressor failure or complete system breakdown.

The substance inside a drier that absorbs moisture is called a desiccant, and can be made of silica, carbon or alumina. The desiccant is very sensitive to moisture. The factory seal on the drier should not be removed until the drier is ready for installation. Do not reuse a filter/drier. Along with trapping moisture, the desiccant will also hold oil and acids. Filter/driers should be replaced whenever a system is opened for service. When removing a filter drier from a system, do not use a torch, use a tubing cutter to avoid releasing moisture into the system. *Never install a suction line drier in the liquid line.*

### *R–410A Considerations*

Liquid line filter driers must have rated working pressures of no less than 600 psig and must be approved for use with R–410A. The technician must always check with the system manufacturer for specific drier recommendations if unsure of what filter drier to use.

## Liquid Line

The liquid line transports high pressure subcooled liquid to the metering device. In transport, the liquid may either lose or gain subcooling depending on the surrounding temperature. Liquid lines may be wrapped around suction lines to help them gain more subcooling, **figure 2–5**. Liquid/suction line heat exchangers can be purchased and installed in existing systems to gain subcooling.

### *R–410A Considerations*

Liquid lines used with R–22 may be used with R–410A if sized correctly and cleaned properly.

**Fig. 2–5. Liquid/suction line heat exchanger (Courtesy, Refrigeration Research, Inc.)**

## Metering Device

The metering device meters liquid refrigerant from the liquid line to the evaporator. There are several different styles and kinds of metering devices on the market with different functions. Some metering devices control evaporator superheat and pressure, and some even have pressure limiting devices to protect compressors at heavy loads. There are five primary types of metering devices;

- ♦ **Thermostatic Expansion Valve**
- ♦ **Automatic Expansion Valve**
- ♦ **Electronic**
- ♦ **Capillary Tube**
- ♦ **Fixed Orifice**

The metering device is a restriction that separates the high pressure side from the low pressure side in a refrigeration system. The compressor and the metering device are the two components that separate pressures in a refrigeration system. The restriction in the metering device causes liquid refrigerant to flash to a lower temperature in the evaporator because of its lower pressure and temperatures.

### *R–410A Considerations*

Metering device capacities increase as the pressure differences across their orifices increase. If the same metering device was used in a R–410A system and a R–22 system, the metering device would be oversized in the R–410A system. This happens because the higher pressures of R–410A systems make for greater refrigerant mass flow rates through the metering device. This is why the flow area in R–410A metering devices are designed to be about 15% smaller than in R–22 systems to achieve the same capacity (tonnage). *Metering device for R–410A and R–22 systems are not interchangeable.*

## Evaporator

The evaporator, like the condenser, acts as a heat exchanger. Heat gains from the product load and outside ambient travel through the sidewalls of the evaporator to vaporize any liquid refrigerant. The pressure drop through the metering device causes vaporization of some of the refrigerant and causes lower saturation temperature in the evaporator. This temperature difference between the lower pressure refrigerant and the product load is the driving potential for heat transfer to take place.

The last pass of the evaporator coil acts as a superheater to ensure all liquid refrigerant has been vaporized. This protects the compressor from any liquid which may result in valve damage or diluted oil in the crankcase. The amount of superheat in the evaporator is usually controlled by a thermostatic expansion type of metering device.

### *R–410A Considerations*

The evaporator or indoor coil should be removed when changing out existing equipment and be replaced with a R–410A specific model. Although some R–22 indoor coils meet the UL approved design and service pressure rating of 235 psig., (heat pump applications), always confirm with the equipment manufacturer before using R–22 indoor coils with R–410A.

## Suction Line

The suction line transports low pressure superheated vapor from the evaporator to the compressor. There may be other components in the suction line such as suction accumulators, crankcase regulators, p–traps, filters, and screens. Liquid/suction line heat exchangers are often mounted in the suction line to transfer heat away from the liquid line (subcool) and into the suction line, **figure 2–6**.

Another function of the suction line is to superheat the vapor as it approaches the compressor. Even though suction lines are usually insulated, sensible heat still penetrates the line and adds more superheat.

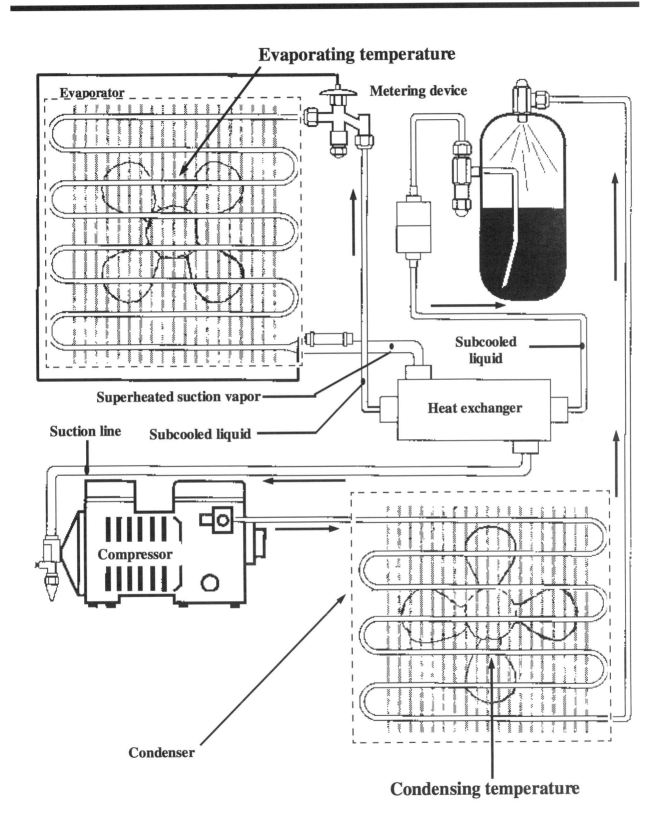

**Fig. 2-6. Refrigeration system showing liquid/suction line heat exchanger.**

This additional superheat decreases the density of the refrigerant vapor to prevent compressor overload, resulting in lower amp draws. This additional superheat also helps ensure that the compressor will see vapor only under low loading conditions. Many metering devices have a tendency to lose control of evaporator superheat at low loads. It is recommended that systems should have at least 20° F of total superheat at the compressor to prevent liquid slugging and /or flooding of the compressor at low loadings.

## R–410A Considerations

Suction lines used with R–22 can also be used with R–410A providing they are correctly sized and properly cleaned.

*Always make sure all components such as reversing valves, expansion valves and filter driers are specifically designed for R–410A.*

# Refrigerant Chemistry and Applications

## Objectives

After completing this section you will be able to:

♦ Compare/contrast the difference in chemical structure of CFC, HCFC, and HFC refrigerants.

♦ Describe a blended refrigerant.

♦ Describe blend fractionation and temperature glide.

♦ Calculate Superheat and Subcooling with blended refrigerants.

♦ Recommend a suitable replacement refrigerant to HCFC–22.

♦ List the basic service tools.

♦ Evaluate the tool requirements for service of the higher pressure refrigerant R-410A

♦ Describe the proper charging procedures for alternative refrigerants.

A refrigerant can be defined as any solid, liquid, or vapor that acts as a cooling agent by absorbing heat from another body or substance. Today, there is no one refrigerant that can be considered the "ideal" refrigerant. The many diverse cooling requirements and applications prevent an ideal refrigerant from existing.

## CFCs, HCFCs, and HFCs

The way in which refrigerants are chemically structured has led to the use of acronyms when referring to the refrigerants. These acronyms are CFCs, HCFCs, and HFCs. All three of these refrigerant acronyms are explained in the following pages.

Most refrigerants contain substances called halogens. Halogen substances are fluorine, chlorine, iodine, and bromine. When combined with a hydrocarbon such as acetylene, methane, or ethane, they are called halogenated refrigerants. The mixture of these chemicals must be precise and form an entirely new substance that has a specific boiling point (saturation temperature). Halide refrigerants are categorized in three groups, ***chlorofluorocarbons, hydrochlorofluorocarbons,*** and ***hydrofluorocarbons,*** according to their chemical makeup.

In 1956, the DuPont Company developed a refrigerant numbering system to identify their refrigerants. At that time, several other companies were manufacturing refrigerants under different trademarks. For example, DuPont's trade name for dichlorodifluoromethane was FREON–12, while Allied Chemicals used GENTRON–12, and Virginia Chemicals used ISOTRON–12.

The American Society of Refrigerating and Air Conditioning Engineers (ASHRAE) has standardized refrigerant identification by using DuPont's numbering system, but precedes each number with the letter "R" (for refrigerant) regardless of the manufacturer. Because of recent concern about ozone depletion, some of the industry has dropped the letter "R" in favor of letters describing the chemical makeup of the refrigerant; (i.e. CFC–12, HCFC–22, HFC–134a, etc.).

Each refrigerant has unique physical properties that make it suitable for a particular application. For example, residential air conditioning systems will generally use HCFC–22, a domestic refrigerator/freezer usually uses CFC–12 (newer models use HFC–134a), and a low temperature commercial freezer may contain R–502 (newer models use HFC–404A).

The EPA has classified R–502 as a CFC, but it is actually an azeotropic mixture of HCFC–22 (48.8%) and CFC–115 (51.2%). When a refrigerant is mixed as an azeotrope, a new refrigerant with unique characteristics is created. The mixture is produced during the manufacturing process under precise conditions and will not fractionalize (separate) if a leak should occur. Azeotropic blends must maintain a specific saturation temperature and strictly conform to Charles's Law of Gases. Service technicians should **never** attempt to mix refrigerants. If the mixture and conditions are not precise, the result will not be azeotropic and will perform according to Dalton's Law of Gases.

ASHRAE has also developed color codes for refrigerant cylinders for ease of identification and to help prevent accidental mixing of refrigerants. Color codes for some of the more common refrigerants are:

| | |
|---|---|
| CFC–11 | Orange |
| CFC–12 | White |
| HCFC–22 | Green |
| HFC–134a | Light blue |
| R–502 | Orchid |
| R–410A | Rose |
| R–407C | Med. Brown |

### Chlorofluorocarbons (CFCs)

CFC's are refrigerants made of chlorine, fluorine and the hydrocarbon (methane). R–11 (trichloromonofluoromethane) and R–12 (dichlorodifluoromethane) are examples of CFC refrigerants. The prefixes (mono = 1, di = 2, and tri = 3) in the chemical names describe how many parts of each element is used in the compound. R–12 therefore contains two atoms of chlorine, two atoms of fluorine, and one carbon atom.

The chlorine in CFC refrigerants will destroy the earth's protective ozone layer when they reach the stratosphere. Because CFC refrigerants are very stable and do not mix with water, they do not break up in the atmosphere like other less stable refrigerants. Because of the environmental impact of CFC refrigerants, the Clean Air Act has promotes the recovery and recycling of refrigerants and has banned the production import of CFC's beginning in 1996.

## *Hydrochlorofluorocarbons (HCFCs)*

Unlike CFCs, HCFCs contain hydrogen atoms that make the compound less stable in the atmosphere. They will break down in the lower atmosphere which makes them less damaging to the ozone layer. Although HCFCs contain chlorine, they have only 2% to 5% of the ozone depletion potential of CFCs.

## *Hydrofluorocarbons (HFCs)*

HFCs contain no chlorine, they have an ozone depletion potential of zero. However, they do have small global warming potentials. Several HFC refrigerants which have increased in popularity are HFC–32, 125, 134a, 143a, 152a, 404A, 407C, and 410A. Therefore, the EPA has mandated recovery of all alternative refrigerants effective November 15, 1995.

## Blends

In the meantime, widespread research is being done for a "drop–in" replacement for CFC–11, CFC–12, CFC–502, HCFC–22 and many other refrigerants. Near azeotropic refrigerant mixtures (NARMs) are now being researched and manufactured by major chemical companies. NARMs, or blends of refrigerants, are being mathematically modeled with computers to tailor the refrigerant blend's characteristics to give maximum system efficiency and performance.

407 c – blend

Refrigerant blends can be HCFC based, HFC based, or a combination of both. Most refrigerant blends are either binary or ternary blends. Binary blends consist of two refrigerants mixed together, while ternary blends consist of three refrigerants. The HCFC based blends are only interim CFC replacements because of their chlorine content. Because HCFCs constitute a major percentage of some blends, these blends have lower ozone depletion and global warming potentials than most CFC and HCFC refrigerants that they are replacing. The HFC based blends will be long–term replacements for certain CFCs and HCFCs until researchers can find pure compounds to replace them.

## Blend Fractionation

Another important phenomenon of near azeotropic and zeotropic refrigerant blends is fractionation. Fractionation is when one or more refrigerants of the same blend may leak at a faster rate than other refrigerants in the blend. Fractionation is a change in composition of the blend by preferential evaporation of the more volatile components, or condensation of the less volatile components. Liquid and vapor must exist simultaneously for fractionation to occur. This different leakage rate is caused from the different partial pressures of each constituent in the near–azeotropic mixture. Fractionation also occurs because the blends are near–azeotropic mixtures and not pure compounds, or pure substances like CFC–12. Fractionation was initially thought of as a serviceability barrier because the original refrigerant composition of the blends' constituents may change over time from leaks and recharging. Depending on the blends' constituent make–up, fractionation may also segregate the blend to a flammable mixture if one or two constituents in the blend is flammable. When leaked, refrigerant blend fractionation may also result in faster capacity losses than single component pure compounds like CFC–12 or HCFC–22. However, further research proved that most blends were near–azeotropic enough for fractionation to be managed without flammability problems.

To avoid fractionation, charging of a refrigeration system incorporating a near–azeotropic blend should be done with liquid refrigerant whenever possible. To ensure that the proper blend composition is charged into the system, it is important that liquid only be removed from the charging cylinder. Cylinders containing near–azeotropic blends are equipped with dip tubes, allowing liquid to be removed from the cylinder in the upright position. Once removed from the cylinder, these blends can be charged to the system as vapor, as long as all of the refrigerant removed is transferred to the system. When adding liquid refrigerant to an operating system (**figure 3–5**), make sure liquid is throttled, thus vaporized, into the low side of the system to avoid compressor damage. A throttling valve can be used (**figure 3–3**) to ensure that liquid is converted to vapor prior to entering the system.

## Blend Temperature Glide

Near–azeotropic ternary blends have temperature glides when they evaporate and condense at a single given pressure. A pure compound like CFC–12, boils and condenses at a constant temperature for a given pressure. Since some blends are near–azeotropic, they will have some "temperature glide" or a range of temperatures in which they will boil and condense. The amount of glide will depend on system design and blend makeup.

Temperature glide can range from 0.2 to 16 degrees Fahrenheit. Since the saturated liquid temperature and the saturated vapor temperature for a given pressure are not the same, the refrigerant in the blend with the highest vapor pressure (lowest boiling point) will seek 100 percent saturated vapor before the other refrigerants. Sensible heat will now be gained by this refrigerant while the other refrigerants in the blend are still evaporating. This same phenomenon happens with condensation.

Some systems will not be affected by this temperature glide because it is design dependent. By all means, system design conditions must be evaluated when retrofitting with a blend. Because of the high percentage of HCFC–22 in some blends, the compressor may see higher condensing saturation temperatures and pressures when in operation. Because HCFC–22 has a relatively higher heat of compression when compared to other refrigerants, higher discharge temperature may be experienced.

## Superheat and Subcooling Calculation Methods for Near–Azeotropic Blends

Near–azeotropic blends are mixtures and not pure compounds, and have an associated temperature glide when they evaporate and condense. Temperature glide is nothing but a range of temperatures when evaporating or condensing for a given pressure. Pure compounds like CFC–12, HCFC–22 and HFC–134a, and azeotropic mixtures or blends like CFC–502 and CFC–500 have only one associated temperature as they evaporate and condense for one given pressure. Because of temperature glide, the methods a service technician will use with near–azeotropes to calculate subcooling and superheat will be different than with pure compounds and azeotropic blends.

## Pressure vs. Temperature R–407C

| TEMP | BUBBLE | DEW |
| --- | --- | --- |
| | *subcooling* | *Superheat* |
| –15 | 17.2 | 9.2 |
| –10 | 21.0 | 12.3 |
| –5 | 25.1 | 15.7 |
| 0 | 29.5 | 19.4 |
| 5 | 34.4 | 23.4 |
| 10 | 39.6 | 27.8 |
| 15 | 45.2 | 32.6 |
| 20 | 51.3 | 37.8 |
| 25 | 57.8 | 43.4 |
| 30 | 64.8 | 49.4 |
| 35 | 72.4 | 56.0 |
| 40 | 80.4 | 63.0 |
| 45 | 89.0 | 70.6 |
| 50 | 98.1 | 78.7 |
| 55 | 107.9 | 87.4 |
| 60 | 118.2 | 96.7 |
| 65 | 129.2 | 106.6 |
| 70 | 140.9 | 117.1 |
| 75 | 153.2 | 128.4 |
| 80 | 166.2 | 140.4 |
| 85 | 180.0 | 153.1 |
| 90 | 194.6 | 166.5 |
| 95 | 209.9 | 180.8 |
| 100 | 226.0 | 195.9 |
| 105 | 243.0 | 211.9 |
| 110 | 260.8 | 228.7 |
| 115 | 279.5 | 246.5 |
| 120 | 299.0 | 265.3 |
| 125 | 319.6 | 285.0 |
| 130 | 341.0 | 305.8 |
| 135 | 363.4 | 327.6 |
| 140 | 386.9 | 350.5 |
| 145 | 411.3 | 374.6 |
| 150 | 436.8 | 399.8 |

## Subcooling & Superheat with Temperature Glide

Manufacturers have now devised pressure/ temperature charts to where it is nearly impossible to choose the wrong temperature for a given pressure. This is because when technicians are figuring superheat values, the chart instructs them to use the **DEW POINT** values only. When technicians are determining subcooling amounts, the chart instructs them to use **BUBBLE POINT** values only. (See table 3–1)

(Table 3–1)

## Evaporator Superheat Calculation

Referring to the pressure/temperature relationship table for R–407C, one can see that there are two temperatures (Dew Point and Bubble Point) involved for one pressure (Bubble–Liquid and Dew–Vapor). Pure compounds such as CFC–12 and azeotropic refrigerant blends have only one temperature for both liquid and vapor phases at a given pressure. In order to calculate evaporator superheat you will need;

- ◆ A pressure temperature chart
- ◆ A gauge manifold set
- ◆ An accurate thermometer

1. Operate the system and note the low side gauge pressure reading.
2. Using an accurate thermometer, determine the evaporator outlet temperature.
3. Using the Dew Point temperature column of the chart, convert the obtained pressure reading to temperature.
4. Deduct the Dew Point temperature from the evaporator outlet temperature.

**Evaporator superheat** is represented in degrees Fahrenheit by the resulting number.

### Example:

| | |
|---|---:|
| Evaporator outlet temperature = | 45° |
| 63.0 Psig low side pressure has a Dew Point conversion of | 40° |
| Evaporator superheat = | 5° |

## Condenser Subcooling Calculations

In order to calculate condenser subcooling you will need;

- ◆ A pressure temperature chart
- ◆ A gauge manifold set
- ◆ An accurate thermometer

1. Operate the system and note the high side gauge reading.
2. Using an accurate thermometer determine the condenser outlet temperature.
3. Using the Bubble Point temperature column of the chart, convert the obtained pressure reading to temperature
4. Deduct the condenser outlet temperature from the Bubble Point temperature.

**Condenser subcooling** is represented in degrees Fahrenheit by the resulting number.

### Example:

| | |
|---|---:|
| Condenser outlet temperature = | 100° |
| 243.0 Psig high side pressure has a Bubble Point conversion of | 105° |
| Condenser subcooling = | 5° |

# ALTERNATIVE  REFRIGERANTS
Commercial and Residential Long Term Replacements.

| ASHRAE# | Trade Name | Manufacturer | Replaces | Type | Lubricant | Application | Comments |
|---|---|---|---|---|---|---|---|
| **R–123** | HCFC–123 | Honeywell | CFC–11 | Pure Compound | Alkylbenzene or Mineral Oil | Centrifugal Chillers | Lower capacity than R–11. With modification equivalent Performance to |
| | | DuPont | | | | | |
| **R134a** | HFC–134a | Honeywell DuPont Elf Atochem ICI | CFC–12 | Pure Compound | Polyol Ester (POE) | New Equipment & Retrofits | Close match to CFC–12 |
| | | | HCFC–22 | | | New Equipment | Lower capacity than HCFC–22. Requires larger |
| **R410A** (32/125) | AZ–20 9100 | Honeywell DuPont | HCFC–22 | Near Azeotropic | Polyol Ester (POE) | New Equipment | Higher efficiency than HCFC–22 Requires |
| **R407C** (32/125/134a) | 407C | Honeywell Elf Atochem ICI | HCFC–22 | Blend (High Glide) | Polyol Ester (POE) | New Equipment & Retrofits | Lower efficiency than HCFC–22, close capacity to |

**Table  3–2**

| Temp | Pressure (psig) | |
|---|---|---|
| F° | R–410A | R–22 |
| 0 | 48.6 | 23.9 |
| 20 | 78.3 | 43.0 |
| 40 | 118 | 68.5 |
| 60 | 170 | 102 |
| 80 | 235 | 144 |
| 100 | 317 | 196 |
| 120 | 418 | 260 |
| 140 | 539 | 337 |

**Table 3–3**

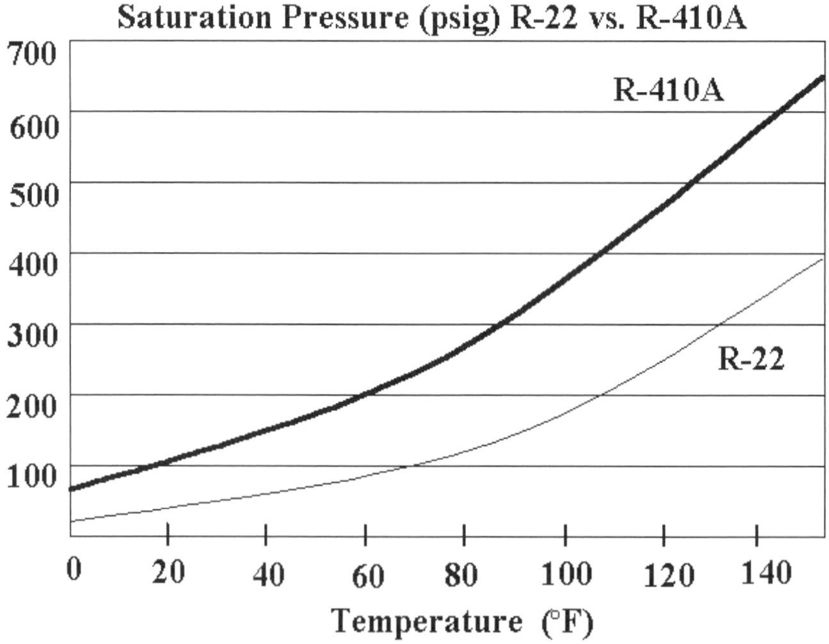

Saturation Pressure (psig) R-22 vs. R-410A

**Figure  3–1**

## Blend Lubricants

The main lubricant for the HCFC–based blends will be a synthetic oil called alkylbenzene. One of the more popular alkylbenzenes has been marketed under the trade name of "Zerol." The blends are soluble in a mixture of alkylbenzene and mineral oils in concentrations of up to 20 percent mineral oil. This will make retrofitting a CFC/mineral oil system to a blend/alkylbenzene system possible without extensive oil flushing. The only system change may be a quick oil flush and a different filter drier. In many of the retrofitted applications, the same thermostatic expansion device will be permitted. Some compressor manufacturers are using a mixture of mineral oil and alkylbenzene lubricants in their compressors for blend applications. Different applications and designs will dictate what lubricants to incorporate in each system. **Most HFC–based blends like R–404A, 407C and 410A will incorporate Polyol Ester lubricants.** Retrofit guidelines have been written, and the original equipment manufacturer should be contacted before retrofitting to a blend.

## HCFC–22 Replacement Candidates

Chemical companies have been researching refrigerants and refrigerant blends in order to find a permanent substitute for HCFC–22. HCFC–22 is scheduled for a total phase–out in the year 2020 under the Montreal Protocol, with partial phase out starting sooner. Listed below are some of the HCFC–22 replacement blends.

## *R–410A*

Refrigerant R–410A is a near–azeotropic, HFC based, binary refrigerant blend consisting of HFC–32 and HFC–125. **However, because R–410A has a very small temperature glide and fractionation potential, the blend is often referred to as an azeotropic blend because it acts much like a single component or pure compound refrigerant.** Its cylinder color is Rose. Its chemical name is Difluoromethane, Pentafluoroethane. It is the leading long–term, non–ozone depleting replacement refrigerant for HCFC–22 (R–22) in "new" residential and light commercial equipment.

R–410A systems operate at much higher pressures (approximately 40% to 70% higher) than standard HCFC–22 systems. In fact, R–22 service equipment (hoses, manifold gauge sets, and recovery equipment) cannot be used on R–410A systems because of these higher operating pressures. Service equipment used for R–410A must be rated to handle higher operating pressures. Safety glasses and gloves should always be worn when working with R–410A. R–410A systems use synthetic Polyol Ester (POE) lubricant in their crankcases, and have higher efficiencies than standard R–22 systems. The **table 3–2** lists other HCFC–22 long–term and interim replacement refrigerants and blends.

R–410A is better known to technicians by such trade names as Suva 410A by DuPont, Puron by Carrier, or Genetron AZ–20 by Allied Signal to name a few. However, all of these refrigerants carry the same American Society of Heating, Refrigeration, and Air Conditioning Engineers (ASHRAE) number of R–410A. Several major U.S. air conditioning original equipment manufacturers (OEMs) using R–410A include Bryant, Carrier, Lennox, Rheem, Ruud, Unico, and Weatherking. An important note to remember is that R–410A is recommended to be used only on new original equipment. It is not recommended as a retrofit to existing R–22 air conditioning systems due to significantly higher operating pressures and higher capacities (**figure 3–3**). In situations where retrofitting would need to be performed on an R–22 system, it is strongly recommended to use R–407C because of it's similar properties to R–22. The higher operating pressures of R–410A have required equipment redesign and some new service tools.

### *Typical Operating Pressures*

Typical air conditioning system operating temperatures may be a 45 degree F. evaporator and a 120 degree F. condenser. The corresponding pressures would then be:

R–410A....................130 psig evaporating pressure
                    418 psig condensing pressure

R–22..........................76 psig evaporating pressure
                    260 psig condensing pressure

### *R–410A Temperature Glide and Fractionation*

Technically, R–410A is a 400 series blend and is classified as a near–azeotropic refrigerant blend. However, because R–410A has a very small temperature glide and fractionation potential, R–410A is often referred to as an azeotropic blend because it acts much like a single–component or pure compound refrigerant such as R–22. In fact, because the temperature glide for R–410A is so small, it is negligible and can be ignored for air conditioning service purposes. The temperature glide for R–410A is less than 0.3 degrees Fahrenheit over air conditioning and refrigerating operating ranges.

### R–410A Pressure/Temperature Chart

Because R–410A has such a low temperature glide, a standard pressure/temperature (P/T) chart can be used when calculating superheat and/or subcooling. There is no need for the type of chart that lists both dew point and bubble point for a specific pressure. Using a standard P/T chart makes things much easier for the technician when servicing R–410A air conditioning systems. (**See table 3–4**)

**(Table 3–4)**

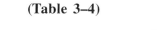

| Pressure vs. Temperature R–410A | |
|---|---|
| **TEMP (°F)** | **PSIG** |
| –15 | 31.3 |
| –10 | 36.5 |
| –5 | 42.2 |
| 0 | 48.4 |
| 5 | 55.1 |
| 10 | 62.4 |
| 15 | 70.2 |
| 20 | 78.5 |
| 25 | 87.5 |
| 30 | 97.2 |
| 35 | 107.5 |
| 40 | 118.5 |
| 45 | 130.2 |
| 50 | 142.7 |
| 55 | 156.0 |
| 60 | 170.1 |
| 65 | 185.1 |
| 70 | 201.0 |
| 75 | 271.8 |
| 80 | 235.6 |
| 85 | 254.5 |
| 90 | 274.3 |
| 95 | 295.3 |
| 100 | 317.4 |
| 105 | 340.6 |
| 110 | 365.1 |
| 115 | 390.9 |
| 120 | 418.0 |
| 125 | 446.5 |
| 130 | 476.5 |
| 135 | 508.0 |
| 140 | 541.2 |
| 145 | 576.0 |
| 150 | 612.8 |

### R–407C

R–407C is an near–azeotropic, HFC–based, ternary refrigerant blend consisting of HFC–32, HFC–125, and HFC–134a with weight percentages of 23%, 25%, 52% respectively. Its cylinder color code is chocolate. Its chemical name is Difluoromethane, Pentafluoroethane, 1,1,1,2–Tetrafluoroethane. R–407C pressures and temperatures are somewhat similar to that of R–22, but it has a slightly lower efficiency than R–22. R–407C is also a non–ozone depleting, long–term replacement refrigerant blend for R–22 in residential and commercial air conditioning and refrigeration applications. R–407C is being used by original equipment manufacturers in new equipment and can also be used as a retrofit refrigerant blend replacing R–22. R–407C has a Global Warming Potential of 0.34. Refer to **table 3–5** for a physical property comparison of R–407C, R–410A, and R–134a.

*less eff. than 22*

*Retrofit for 22*

R–407C is classified as a near–azeotropic refrigerant blend and does not act like a single component or pure compound refrigerant when evaporating and condensing. R–407C has a large temperature glide (9–12 degrees F) and fractionation potential over air conditioning and refrigeration temperatures ranges. These qualities cannot be ignored when the service technician is working with R–407C. R–407C performs similar to R–22 under evaporator temperatures ranging from 20 to 50 degrees F.

### R–407C Temperature Glide and Fractionation

Because R–407C has such a high temperature glide, the technician will have to use a pressure/temperature (P/T) chart like the one in (**table 3–1**). Notice that there is both a **DEW POINT** temperature and a **BUBBLE POINT** temperature for each pressure value listed for R–407C. When the technician is figuring a superheat value, the chart allows them to use the DEW POINT temperature values only. When figuring subcooling amounts, the chart allows them to use BUBBLE POINT temperature values only. This type of P/T chart, which was devised by manufacturers, makes it nearly impossible for service technicians to use the wrong temperature for a given pressure when figuring superheat and subcooling values.

| Physical Properties of Refrigerants | R–407C | R–410A | R–22 | R–134a |
|---|---|---|---|---|
| Composition | R–32/R–125/R134a | R–32 / R–125 | Single Component | Single Component |
| (weight %) | (23 / 25 / 52) | (50 / 50) | Not Applicable | Not Applicable |
| Molecular Weight | 86.2 | 72.6 | 86.5 | 102.3 |
| Boiling Point (1 atm, F) | –43.6 | –61 | –41.5 | –14.9 |
| Critical Pressure (psia) | 672.1 | 691.8 | 723.7 | 588.3 |
| Critical Temperature (F) | 187 | 158.3 | 205.1 | 213.8 |
| Critical Density (lb./ft^3) | 32 | 34.5 | 32.7 | 32.04 |
| Liquid Density (20 F, lb./ft^3) | 72.35 | 67.74 | 75.27 | 76.21 |
| Vapor Density (bp, lb./ft^3) | 0.289 | 0.261 | 0.294 | 0.328 |
| Heat of Vaporization (bp, BTU/lb.) | 106.7 | 116.8 | 100.5 | 93.3 |
| Specific Heat Liquid (20 F, BTU/lb. F) | 0.3597 | 0.3948 | 0.2967 | 0.3366 |
| Specific Heat Vapor (1 atm, 20 F, BTU/lb. F) | 0.1987 | 0.1953 | 0.1573 | 0.2021 |
| Ozone Depletion Potential (CFC–11 = 1.0) | 0.0 | 0.0 | 0.05 | 0.0 |
| Global Warming Potential (CFC–11=1.0) | 0.34 | 0.39 | 0.35 | 0.28 |
| ASHRAE Standard 34 Safety Rating | A1/A1 | A1/A1 | A1 | A1 |
| Temperature Glide (see Section Two B) | 10 | 0.2 | Not Applicable | Not Applicable |

**Table 3–5**

# Basic Service Tools

## Gauge Manifold

One of the most important tools a service technician uses is the gauge manifold. The gauge manifold is a pressure checking device with both compound and high pressure gauges. A compound gauge can measure pressures above and below atmospheric (pressure and vacuum).

The saturation temperatures of the most popular refrigerants are usually included on the face of the gauge dial. Remember that the gauges will only reveal the saturation temperature at a given pressure. The gauges will not reveal superheat unless a thermometer is used to compare the actual temperature to the saturation temperature.

Because of changes in atmospheric pressure, it is not unusual for the gauges to require recalibration in the field. To gain access to the calibration screw, remove the clear cover from the gauge. The calibration screw is usually on the gauge face, just below the pointer hub. While the gauge is exposed to atmospheric pressure, turn the calibration screw slowly until the pointer lines up with 0 psig.

The manifold has control valves and connections for hoses to the service valves. Some manifolds have two valves (low and high side), while others have four (two extra valves for vacuum pump and refrigerant charging cylinder connections).

## R–410A Considerations

The gauge manifold set is specially designed to withstand the higher pressures associated with R–410A. Manifold sets are required to range up to 800 psig on the high side and 250 psig on the low side with a 550 psig low side retard.

All hoses should have a service rating of 800 psig. (This information will be indicated on most hoses.)

## Micron Gauge

A micron gauge must be used when evacuating a system to 500 microns as the manifold gauges will not read accurately the deeper vacuum measurements.

## Vacuum Pumps

An evacuation to 500 microns is usually sufficient to remove moisture from a system using R–22 and mineral oil lubricant.

## R–410A Considerations

A 500 micron evacuation, however, will not separate moisture from Polyolester oil (POE) in R–410A systems. In addition to a 500 micron evacuation, a liquid line filter drier must be installed. The liquid line filter drier (must be R–410A compatible) should be replaced any time the system is opened.

*its okay to oversize*

## Leak Detectors

An electronic leak detector capable of detecting HFC refrigerant can be used with R–410A refrigerant. Older R–22 leak detectors, as well as halide torch leak detectors will not detect leaks in R–410A systems. Never use air and R–410A to leak check, as the mixture may become flammable at pressures above 1 atmosphere. A system can be safely leak checked by using nitrogen or a trace gas of R–410A and nitrogen. (Remember always use a pressure regulator with nitrogen and a safety relief valve down stream set at no more than 150 psi.)

Bubble solutions are commercially available specifically for leak checking and will work with R–410A refrigerant.

There are six (6) main types of detectors that can be used for monitoring alternative refrigerant leaks.

1. **Non–Selective Detectors**
2. **Halogen Specific**
3. **Compound Specific**
4. **Infrared–Based**
5. **Fluorescent Dyes**
6. **Ultrasonic**

## R–410A Considerations

If the R–410A system develops a leak, the technician does not have to recover the remaining refrigerant from the system before they "top–off" the system. Because R–410A is close to being an azeotropic blend, it behaves like a pure compound or single component refrigerant. **The technician can use the existing refrigerant in the system after leaks have occurred. There is no significant change in the refrigerant composition during multiple leaks and recharges. However, the service technician must remember that when adding R–410A to the system, it must come out of the charging cylinder as a liquid to avoid any fractionation and for optimum system performance.** If the air conditioning system has lost its complete charge, the system should be leak checked, repaired, and evacuated to below 500 microns. A digital scale or a calibrated charging cylinder designed for the greater pressures of R–410A should then be used to recharge R–410A back into the system.

The Refrigerant Recycling Regulations of Section 608 of the Clean Air Act Amendments state that technicians must find and repair "substantial leaks" for systems with 50 pounds or more of refrigerant.

***Substantial leaks are:***
♦ 35% annual leak rate for commercial and industrial refrigeration
♦ 15% annual leak rate for comfort cooling chillers and all other equipment

***Never use air to leak check any refrigerant system. Mixtures of air and R–410A or R–22 can cause explosive mixtures at certain concentrations and pressures.***

## Refrigerant Recovery Systems

Removal of refrigerant from a system can be accomplished by two basic methods, passive and active. To comply with government regulations and best serve customer needs, time must be taken to evaluate the system and determine which method to employ.

Questions to be considered by the technician.
♦ Is the system compressor operable?
♦ Is the system accessible enough?
♦ Where is the liquid refrigerant within the system?
♦ What is the outside (ambient) temperature?
♦ Will outside conditions have any effect?

If the system is not analyzed, recovery could take longer than necessary.

### *Passive Recovery (System–Dependent)*

System–Dependent recovery or "Passive" recovery, is recovering refrigerant from a system employing the refrigeration system's internal pressure and/or system's compressor as an aid in the recovery process. System–Dependent equipment can not be used with appliances containing more than 15 pounds of refrigerant. To make it easier for technicians to recover refrigerant, the EPA is requiring the manufacturer to install a service aperture or process stub, for appliances containing Class I and II refrigerants. If a service technician uses "Passive or System Dependent recovery on a system with an inoperative compressor, the refrigerant must be recovered from both the low and high side of the appliance to speed the recovery process and to achieve the required recovery efficiency requirements. A vacuum pump can be used in this procedure, however, ***never discharge a vacuum pump into a pressurized container.*** Vacuum pumps cannot handle pumping against anything but atmospheric pressure. If the compressor is operative, refrigerant can be recovered from the high side only. In all passive recovery, the refrigerant must be recovered in a non–pressurized container. Whether the compressor is operative or not, gently striking the compressor with a wood or rubber mallet during recovery will agitate and release the refrigerant dissolved in the compressor's crankcase oil. Contingent upon the following, refrigerant can be removed without damage to the compressor.

♦ An adequately sized receiver or condenser.
♦ Weight recording method.
♦ Proper on–off controls.
♦ Adequate recovery containers not exceeding container's maximum net weight

## Active Recovery (Self-Contained)

The most common method of system refrigerant removal is through use of a certified self–contained recovery unit. Self–contained (active) recovery equipment has its own means of removing refrigerant from appliances and is capable of reaching the required recovery rates whether or not the appliance compressor is operable. Self contained recovery equipment stores refrigerant in a pressurized recovery tank. Before operating a self–contained recovery machine, make sure that the tank inlet valve is open, and that the recovery tank does not contain excessive non-condensables, (air). Follow the operating instructions supplied by the recovery equipment manufacturer regarding purging of non–condensables. All refrigerant recovery equipment should be checked for oil level and refrigerant leaks on a daily basis. Some machines are capable of both liquid and vapor removal. A higher ambient temperature facilitates more rapid recovery due to increased system internal vapor pressure.

## R–410A Considerations

Due to the higher pressure of R–410A (50–70% higher than R–22). component and servicing equipment have been redesigned to withstand the increased pressure. Recovery and recycling equipment rated for the higher R–410A pressures, must be used. (Consult manufacturers for proper equipment recommendations.) Recovery cylinders must have a service rating of 400 psig (DOT 4BA 400 and DOT 4BW 400 are acceptable cylinders.) ***Do not use standard DOT recovery or storage cylinders rated at 300 psig with R–410A.***

As stated previously manifold gauges used with R–410A require a high side range of 800 psig and a low side of 250 psig with a 550 psig low side retard. Hoses are required to have a service pressure rating of 800 psig.

Avoid the mixing of R–410A with other refrigerants during the recovery and recycling of system refrigerants. To prevent the mixing of refrigerants, sometimes called cross contamination, the technician should use a self–clearing or self–purging recovery/recycle unit. Manifold gauges, hoses and recovery cylinders should be evacuated after every recovery job.

Another method that will eliminate cross contamination is to dedicate equipment to R–410A systems. All dedicated equipment should be marked clearly for R–410A use only. This would include:

- R–410A recovery/recycle unit
- R–410A manifold gauge and hoses
- DOT 4BA 400 or DOT 4BW 400 recovery cylinders
- A deep vacuum pump capable of 500 micron pull down
- A manual or automatic solenoid shut–off scale

## Refrigerant Charging

The proper refrigerant charge is necessary to insure the equipment is operating at its maximum efficiency and functioning as designed by the manufacturer. Many problems occur if the system is undercharged or overcharged.

### *Undercharge*

An undercharge can cause excessive flash gas to enter the metering device creating the following:
1. Low evaporator temperature.
2. Excessive superheat.
3. Underfed evaporators.
4. High compression ratios.

### *Overcharge*

An overcharge can cause the metering device to over feed the evaporator and backup liquid refrigerant in the condenser. These conditions could create the following:
1. Flood back.
2. Liquid slugging.
3. High side pressure increase.
4. Loss of capacity.
5. High compression ratios.

## *R–410A System Charging*

Even though R–410A has a very small fractionation potential, it cannot be ignored completely when charging. To avoid fractionation, charging of an air conditioning system incorporating R–410A should be done with "**liquid**" to maintain optimum system performance. Follow the instruction on the charging cylinder if unsure of the charging procedure. To insure that the proper blend composition is charged into the system, it is important that liquid only be removed from the charging cylinder. Some cylinders supplied by manufacturers have dip tubes which allow liquid refrigerant to be removed from the cylinder when it is in the upright position (**figure 3–2**). Cylinders without dip tubes have to be tipped upside down in order for liquid to be removed. The service technician must differentiate between which type of charging cylinder they are using to avoid removing vapor refrigerant instead of liquid refrigerant to avoid fractionation and for safety concerns.

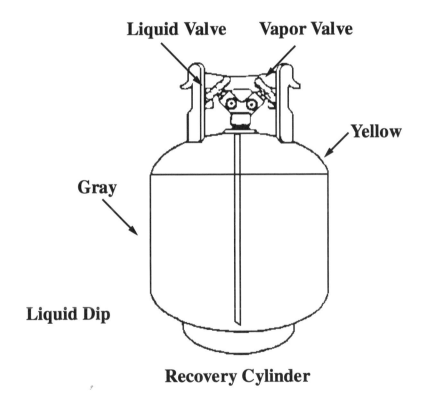

**Figure 3-2**

Once the liquid is removed from the charging cylinder, R–410A can then be charged into the system as vapor as long as all the refrigerant in the charging cylinder is charged into the system. Remember, if the service technician wants to add liquid refrigerant to an operating system, make sure liquid is **throttled**, thus vaporized, into the "low side" of the system to avoid compressor damage. A throttling valve can be used to ensure that liquid is converted to vapor prior to entering the system (**figure 3–3**). Proper manipulation (restricting) of the manifold gauge set can also act as a throttling device to ensure liquid is not entering the compressor.

**Figure 3-3     Throttling Valve**

**Courtesy Thermal Engineering**

Many service technicians will often attempt to charge an air conditioning or refrigeration system by making sure the liquid line sight–glass is full of liquid refrigerant. Often, as a refrigerant blend travels through a liquid line sight–glass, some of the liquid may flash as it passes through the increased volume of the sight–glass (**figure 3–4**). Once the small percentage of flashed liquid leaves the sight–glass and re–enters the smaller liquid line, it will form 100% liquid again. Because of this flashing phenomenon within the sight–glass with certain refrigerant blends, technicians often think the system is undercharged. If the system has a sight glass, it is of utmost importance for service technicians not to try to clear the sight–glass when charging with refrigerant blends like R–410A or R–407C. Attempts to clear a sight–glass may overcharge the system and lead to poor system performance and/or compressor damage.

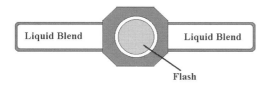

**Figure 3–4 Sight Glass**

## Charging for Proper Subcooling R–410A

If a system uses a thermostatic expansion valve, the device will regulate the refrigerant flow over a wide range of load and charge conditions. Therefore, some manufactures recommend using subcooling to check for proper charge conditions.

*Note: Restrict air flow across condenser and bring condenser pressure to 350 psig if outdoor temperature is less than 65 degrees Fahrenheit.*

1. Operate system for at least 10 minutes to stabilize.
2. Attach gauges to liquid valve port and measure the liquid line pressure. Use pressure/ temperature chart or gauge to determine the saturation temperature that corresponds to that pressure.
3. Measure the liquid–line temperature at the liquid line as close to the outdoor coil as possible using a fast reading temperature probe.
4. The difference between the saturation temperature and the actual liquid–line temperature is the subcooling.
5. Use manufacturers recommendations – If no information is available, use a subcooling value 10–15° F.
6. Make any adjustments by adding refrigerant to increase subcooling and removing refrigerant to decrease subcooling.

This method of charging requires the use of accurate refrigeration gauges, dry bulb thermometer and a pressure/temperature chart or the pressure/temperature conversion face of the gauges.

### Charging for Proper Superheat R–410A

This applies only to fixed metering device systems, such as, fixed orifice, (restrictor) or capillary tube. This method is for cooling only charging, refer to equipment manufacturers instructions for heat pump charging.

**Note:** Restrict air flow across condenser and bring condenser pressure to 350 psig if outdoor temperature is less than 65 degrees Fahrenheit.

1. Operate the system for at least 10 minutes to stabilize.
2. Attach gauges to suction valve port and measure suction pressure. Use pressure temperature chart or gauge to determine the saturation temperature that corresponds to the suction pressure.
3. Measure the suction temperature at the suction line approximately 6" before the compressor inlet using a fast reading temperature probe.
4. The difference between the saturation temperature and the actual suction line temperature is the superheat.
5. Compare calculated superheat with the allowable range of superheat for existing conditions, indicated by the manufacturers specifications.
6. Make any adjustment by:
   Adding refrigerant to lower superheat & removing refrigerant to increase superheat.

### Precautions

♦ Do not vent refrigerant.
♦ Use recovery equipment and cylinders approved for R–410A.
♦ Always charge with liquid, using a commercial metering device in the manifold hose.
♦ If the cylinder has a dip tube, keep the cylinder upright for liquid.
♦ If the cylinder does not have a dip tube, invert the cylinder to obtain liquid.

This method of charging requires the use of accurate refrigerant gauges, a psychrometer, or wet bulb and dry bulb thermometers and a pressure/temperature chart or the pressure/temperature conversion face of the gauges.

## *R–407C System Charging*

R–407C has the ability to fractionate and cause a permanent change in the composition of the refrigerant charge. Because of this, it is recommended to remove R–407C from the charging cylinder as a liquid to ensure maximum system performance. Follow the instruction on the charging cylinder if unsure of the charging procedure. Once the liquid is removed from the charging cylinder, R–407C can be charged into the system as a vapor as long as all the refrigerant removed from the charging cylinder is charged into the system. Remember, when adding liquid refrigerant to an operating system, make sure liquid is **throttled**, thus vaporized, into the low side of the system to avoid compressor damage. The same methods used for charging R–410A systems hold true for R–407C systems, even though R–407C systems have more of a fractionation potential. Since R–407C system pressures and temperatures are somewhat similar to an R–22 system, the same manifold gauge set and charging cylinder types can be used.

## *R–407C Refrigerant Leaks & Leak Detectors*

Since R–407C and R–410A are both HFC–based refrigerant blends, the same methods and procedures hold for both refrigerants on the subject of refrigerant leaks and leak detectors.

# Adding refrigerant to a system while in operation

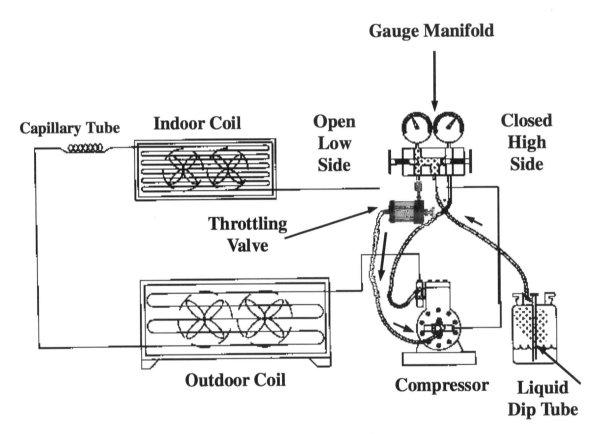

Figure 3-5

# Refrigerant Oils and Their Applications

**4**

## Objectives

♦ Describe the function of refrigeration oil.
♦ Compare / contrast different types of refrigeration oils.
♦ Recommend proper procedures for waste oil disposal.
♦ Select an oil suitable for alternative refrigerants.
♦ Explain the hygroscopic properties of different refrigerant oils.

Today, oils used in refrigeration compressors are not considered standard lubricants by any means. Years of testing and research have made a science out of refrigeration oils thus categorizing them as specialty products. Understanding the behavior of refrigeration oil requires information on composition, properties, and application.

In any refrigeration system, oil and refrigerant are always present. The refrigerant is the working fluid and is required for cooling. The main purpose of the oil is to lubricate the compressor. Refrigerant and oil are miscible (mixable) in one another and their magnitude of miscibility will depend on the type of refrigerant, the temperature, and the pressure both are exposed to. A certain amount of oil will always leave the compressor's crankcase and be circulated with the refrigerant. Refrigerant and oil can separate into two phases and become immiscible in one another at certain temperatures. Refrigeration oils and refrigerants are often miscible in one another over wide temperature ranges. If not soluble, the oil would not move freely around the system and oil rich pockets would form causing restrictions, poor heat transfer, and inadequate oil return to the compressor.

Even though the primary function of an oil is to minimize mechanical wear through lubrication and to reduce the effects of friction, oil in a refrigeration system accomplishes many more tasks. Oil acts as the seal between the discharge and suction sides of the compressor. Oil will prevent excessive blow–by around the piston in a reciprocating compressor. Oil also prevents blow–by in some centrifugal compressor by putting a seal around its vanes. Oil acts as a noise dampener reducing internal mechanical noise within a compressor. Oil also performs heat transfer tasks by sweeping away any heat from internal rotating and stationary parts.

---

**Refrigeration Oil Functions**

Minimizes mechanical wear
Reduces friction
Lubricates
Seals – prevents blow–by
Deadens noise
Transfers heat – "cools"

---

## Oil Groups

As service technicians, it is important to realize the magnitude of the refrigerant and oil transition this industry is experiencing. Refrigerants and oils have become a complex science. There used to be "rules of thumb" to follow that matched a certain viscosity with the temperature application. The diversification of oils and oil additives used with today's ozone friendly refrigerants and even yesterday's refrigerants make these rules of thumb obsolete. Education through reading current literature is one method a technician can use to keep abreast of these new technologies and changes in our industry. A technician can no longer rely on the rule of thumb for adding oil to a system. Technicians must always refer to the manufacturer's literature for each compressor to get information on what oil to incorporate.

Oils in general can be categorized into two groups, Natural and Synthetic.

The natural groups are animal, vegetable, and mineral oils. Both animal and vegetable oils cannot be refined or distilled without a change in composition. Both are considered poor lubricants in the refrigeration industry because of this changing composition. Poor stability is another disadvantage of animal and vegetable oils. These oils will form acids and gums very easily. Another problem with animal and vegetable oils in refrigeration applications is their somewhat fixed viscosity. Different viscosities for the diverse temperature applications in the refrigeration and air conditioning industries are mandatory and cannot be reached with animal or vegetable oils.

## Synthetic Oils

Because of the somewhat limited solubility of mineral oils with certain refrigerants such as R–22, synthetic oils for refrigeration applications have been used successfully. Three of the most popular synthetic oils are alkylbenzenes, glycols, and ester based oils.

## Alkylbenzene

Most HCFC–based refrigerant blends perform best with alkylbenzene lubricants when compared to other mineral oils. This is because existing mineral oils are not completely soluble in the refrigerant blends. The blends are soluble in a mixture of mineral oil and alkylbenzene up to a 20 percent concentration of the mineral oil. This indicates that mineral oil systems that are retrofitted with refrigerant blends will not require extensive flushing of the oil. Today, as in the past, alkylbenzenes are used quite often in refrigeration applications (i.e., Zerol).

## Glycols

Some of the most popular glycol–based lubricants are polyalkylene glycols (PAGs). They were the first generation lubricants used with HFC–134a. However, many polyalkylene glycol lubricants tested with HFC–134a are not fully soluble and will separate. PAG oils also have a track record of being very hygroscopic (attract and retain moisture). However, modified PAGs are still being researched. Another disadvantage of PAG oils is their poor aluminum on steel lubricating abilities. PAGs also have been known to have a reverse solubility in refrigeration systems. This means that the oil may separate in the condenser instead of the evaporator. PAGs also have a very high molecular weight and can be harmful if inhaled in certain concentrations.

## Esters

Other synthetic oils gaining popularity are the ester based oils. A major advantage of ester based oils is their wax free composition. No wax will give a lower pour point. A popular ester–based oil is Polyol Ester. Ester based oils are also used extensively in many HFC–based refrigerant blends like R–410A, R–407C and R–404A. Ester oils are very hygroscopic meaning they will attract and retain moisture easily. However, always consult with the compressor manufacturer before using any lubricant.

## Waste Oil

Although the Environmental Protection Agency (EPA) has ruled that refrigeration oils are not hazardous wastes, disposing of used oils in a careless fashion is against the law.

The EPA has specifically exempted "on condition that these used oils are not mixed with other wastes, that the used oils containing CFCs are subjected to recycling and/or reclamation for further use, and that these used oils are not mixed with used oils from other sources."

Used oil is hazardous if a tested sample is found to contain specific compounds. Concentrations of specific compounds such as mercury, cadmium, or lead, or if the waste exhibits characteristics of ignitibility or corrosiveness, fall under description according to EPA for hazardous waste management.

It remains your responsibility to determine if your waste is hazardous. You are obligated to make sure that the waste, if hazardous, is disposed of safely and legally. Basically, it is your waste. You own it ... Forever.

## Lubricants for *HFC R–410A, R–407C, & R–134a*

Manufacturers of both hermetic and semi–hermetic compressors have determined that Polyolester (POE) lubricants are the best choice for chlorine–free refrigerants. HFC refrigerants will not mix with mineral oils or alkylbenzenes. POE lubricants are completely wax free and have a proven track record for good miscibility (mixing ability) between the oil and HFC refrigerants. This allows the oil to stay in solution with the refrigerant and helps the oil get back to the compressor's crankcase. New R–410A and R–407C systems will be supplied with the proper POE lubricant already in the system.

It is important for the service technician to understand that not all POE lubricants are interchangeable. There are many types and grades of POE lubricants that give them different properties. Avoid mixing POEs from different manufacturers or viscosity grades. If the lubricant has to be added or replaced, the service technicians must consult with the original equipment manufacturer (OEM) if unsure what type or grade of POE is used in a particular air conditioning or refrigeration system. Do not just simply add any POE oil off of the shelf when servicing an air conditioning or refrigeration system because system incompatibilities with the lubricant may exist. POE lubricants are made from more expensive base stock materials than most mineral oils.

## Some important advantages of POE lubricants over mineral oils are listed below:

♦ POEs are miscible with CFC, HCFC, and HFC refrigerants

♦ Better oil return characteristics than mineral oils

♦ Improved heat transfer characteristics than mineral oils

♦ As good or better lubricating ability as mineral oils

♦ POEs are wax free lubricants

## Special Concerns with Polyolester (POE) Lubricants

♦ POE lubricant is hygroscopic (readily absorbs and retains moisture from the atmosphere).

♦ Never store POE in a plastic container, always use a glass or metal container.

♦ Use a pump to transfer POE lubricants.

♦ Use an approved POE lubricant (POE lubricants are not always interchangeable).

♦ A vacuum pump will not remove moisture effectively from the POE lubricant (a liquid line filter drier must always be used).

♦ POEs can be irritating to the skin

♦ POEs can damage some roofing membrane material

♦ POEs are better solvents than mineral oils

POE lubricants are very hygroscopic, they will rapidly absorb moisture and strongly hold this moisture in the oil (**figure 4–1**). Service technicians must realize that they must minimize the lubricants exposure to the atmosphere whenever possible. If the lubricant is exposed to the atmosphere, limit the exposure to no more than 15 minutes. Always store POE oils in metal or glass containers. When stored in plastic containers, the POE oil has the ability to absorb moisture through the plastic container. It is recommended to use a pump to move or transfer the POE lubricant from its container to the refrigeration or air conditioning system. If the system is under a vacuum, break the vacuum with the appropriate refrigerant for the system or with dry nitrogen. Never break the vacuum with air because of the exposure to atmospheric moisture. A liquid line filter drier will remove moisture from POE oils that may get into the system. Vacuum pumps are often incapable of removing moisture from the POE oil. Vacuum pumps are designed for removing free water. Even a very deep vacuum will not pull the water molecules away from the POE lubricant. A combination of these practices will help keep atmospheric moisture exposures to the POE lubricant to a minimum.

POE oils can cause skin irritations. It is recommended to use appropriate gloves and safety glasses when handling POE lubricants. When finished handling the POE lubricant, thoroughly wash with soap and water to remove any residual oil.

POE oils have been known to damage some membrane roofing materials. The service technician should always protect the working area surface when working with POE lubricants.

POE oils are excellent solvents. They are much better solvents than mineral oils. Refrigeration and air conditioning systems with residue on the inside of their piping and components may have this residue washed into other parts of the system such as the compressor or valve arrangements. Liquid and suction line filter driers are recommended in POE systems. There are special driers made for cleaning–up wet POE systems. These special driers should be replaced frequently until system moisture levels are back to normal. A normal filter drier can then be installed for the long term.

## Water Absorption into Lubricants

**Exposure of Lubricant to Moist Air**
**(PPM Water Absorbed over Time)**

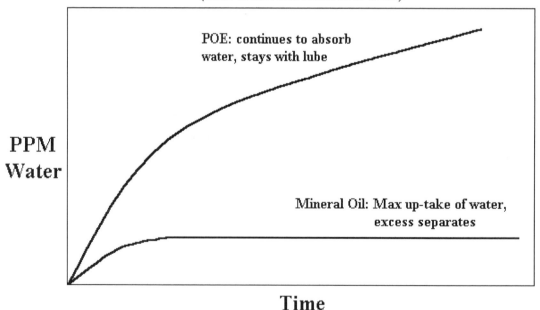

Figure 4–1

# *Safety*

## *Objectives*

After completing this section you will be able to:

- ♦ Describe safe refrigerant handling procedures
- ♦ Evaluate the safety protection policies of the workplace (ASHRAE Standard 15)
- ♦ Identify refrigerant classification according to ASHRAE Standard 34
- ♦ Review all safe service procedures for the higher pressure refrigerant R–410A

## Personal Safety Protection

It is important that technicians receive proper safety training and become familiar with policies that include those of the Occupational Safety and Health Administration (OSHA), other federal and state regulations, and the policies of the company or job site.

Approved eye protection must be worn when working with refrigerants, power tools, or anytime there is danger from flying debris. Ear protection must be worn when working in noisy areas. Avoid loose fitting clothes, wrist watches, rings, etc., when working near machinery. Safety shoes and hard hats are required equipment at many workplaces.

When performing service work, be aware of the nearest emergency exit, fire extinguisher and first–aid station.

Keep the shop area clean. Work areas, isles, exits and stairways need to be free from obstructions. Keep floors clean and free of spilled liquids.

Keep hand and power tools in proper working condition. Always use the right tool for the job.

When use of ladders is necessary, select the correct ladder for the job. Inspect the ladder to sure it is in good condition, free from damage, grease or oil. Be sure that the ladder is set on a firm and level base. When working on a roof tie off the ladder, and be certain that the ladder extends above the roof by three rungs. Do not use metal ladders near power lines.

If a job requires fixed or mobile scaffolding, it must be in compliance with OSHA and other federal standards. Always wear a safety belt when working on ladders or scaffolds.

Nitrogen cylinders are shipped at a pressure of 2500 psig. All nitrogen cylinders should be stored or moved with the protective cap in place. Store cylinders chained to a wall or to a moving cart designed for this purpose. Dropping a cylinder can cause the valve to break off, propelling the cylinder like a missile.

Before using nitrogen for pressure testing, the 2500 psig must be reduced to a safe working pressure (150 psig should suffice, but check the data plate on the unit for the safe test pressure). A pressure regulator and a safety relief valve must be used.  Nitrogen is an inert gas and does not support combustion. **NEVER** pressurize a system with oxygen or compressed air. If there is oil or oil residue in the system, adding oxygen will cause an explosion.

## Electrical Safety

Current is the killing factor in electrical shock. Ohm's Law explains the relationship between voltage, current, and resistance. Human bodies have resistance when voltage is applied, current will flow. If only one tenth of the current required to operate a 10 watt light bulb were to pass through your chest, the results could be lethal. Most people are killed by 110–V power, probably because we all tend to take it for granted. Ohm's Law states that the amount of current passing through a conductor is directly proportional to voltage applied.

If 110 volts were placed across a 500 ohm resistance, the resulting current would be 0.22 amps, or 220 mA. (The "m" stands for milli, or 1/1,000) A current of 2 to 3 mA will generally cause a tingling sensation. The tingling sensation increases and becomes very painful at about 20 mA. Current between 20 and 30 mA will cause muscle contraction and you my be unable to let go of the wire. Currents between 30 and 60 mA will cause muscular paralysis and difficulty breathing. Breathing at 100 mA current is extremely difficult. Currents between 100 and 200 mA generally cause death because the heart goes into fibrillation. A 110–V power circuit will generally cause between 100 and 200 mA current flow through the bodies of most people.

Electrical power should be shut off at the entrance or distribution panel then locked and tagged in an approved manner to prevent accidental activation. The technician who locks–out the power source should keep the only key in their possession. (**See figure 5–1**)

The tag should contain the following information:

- ◆ Technician's name
- ◆ Service being performed
- ◆ Reason for service
- ◆ Date and time

*Specific lock–out and tag requirements are furnished by the Occupational Safety and Health Administration (OSHA).*

When attached to a power source, both lockout and tagout devices used in accordance with OSHA standards help protect employees from hazardous energy. A lockout device provides protection by physically preventing the machine or equipment from becoming energized. A tagout device provides protection by indicating that the equipment being controlled may not be operated while the tagout device is in place. Lockouts or tagouts must be singularly identified, must be the only devices used for controlling hazardous energy, and must meet the following requirements:

***Durability*** – Lockout and tagout devices must be constructed and printed so that they do not deteriorate or become illegible, especially when used in corrosive or wet environments.

***Standardized*** – Lockout and tagout devices must be standardized according to either color, shape, or size, and must also be standardized according to print and format.

***Substantial*** – Locks must be substantial to prevent removal except by excessive force of special tools, such as bolt cutters. Tag attachment must be self–locking, non–reusable, and non–releasable, such as a nylon cable tie that will withstand all environments and conditions.

The procedures for lockout or tagout must include the following steps;
 **1**) prepare the equipment for shutdown,
 **2**) shut down the equipment,
 **3**) isolate the equipment from the energy source,
 **4**) apply the lockout or tagout device to the isolating device,
 **5**) safely release all potentially hazardous stored or residual energy,
 **6**) verify the isolation of the equipment prior to the start of service work.

**Figure 5–1 Sample Tag**

Before lockout or tagout devices are removed and energy is restored to the machines or equipment, certain steps must be taken after servicing is completed, including:

**1**) ensuring that equipment components are operationally intact;

**2**) ensuring that all employees are safely positioned or removed from equipment;

**3**) ensuring that lockout or tagout devices are removed by the employee who applied the device.

Certain tests must be performed with the power on. Exercise extreme caution when testing live circuits. Always know the voltage of the circuit you are working on and act accordingly. Competent technicians always take precautions when working with electrical circuits. Do not work alone. If you must test a live circuit, have someone with you to turn off the power, call for help, or give cardiopulmonary resuscitation (CPR).

## Safe Refrigerant Handling

Refrigerants and pressurized cylinders can be dangerous if not handled properly. Technicians must understand and follow all safety precautions before handling any refrigerant. Technicians should read and understand the material safety data sheet for all oils, refrigerants, any other chemicals they may use in the workplace.

Refrigerants are heavier than air and will displace oxygen in a closed area. Proper ventilation should be in place before working in a confined space. If a refrigerant leak occurs, immediately vacate the area.

## Storage Cylinders

The pressurized refrigerant in a cylinder is potentially dangerous. Always wear safety glasses, protective clothing and gloves when working with refrigerants. A release of high–pressure liquid refrigerant to atmospheric pressure will cause the refrigerant to flash and boil to a vapor, absorbing heat from anything it comes in contact with. If the refrigerant come in contact with the skin or eyes, frostbite or blindness can occur.

R–410A cylinders must be clearly marked and stored in a cool, dry, properly vented storage area away from heat, flames or corrosive chemicals. Store cylinders of R–410A out of direct sunlight.

**NEVER ALLOW A CYLINDER OF R–410A TO GET WARMER THAN 125° F (52°C).**

446 psig

R–410A expands significantly at warmer temperatures, reducing the vapor space in the cylinder. Once a cylinder is full of liquid, any further rise in temperature will cause it to burst.

All storage and shipping containers must be specially designed for R–410A. This includes cylinders, storage tanks, tank trailers or tank cars. Although R–410A has a small fractionation potential, R–410A must be transferred as liquid. Vapor transfers could cause a change in the composition of the refrigerant.

Empty recovery cylinders should be evacuated before transferring refrigerant. **DO NOT MIX REFRIGERANTS**. Mixing refrigerants can cause dangerously high pressures and may be impossible to reclaim.

Refrigerant recovery cylinders must never be filled over 80% of their liquid capacity. This leaves 20% of the volume for expansion. If a cylinder were completely filled, expansion of the refrigerant and the resulting hydrostatic force would cause the cylinder to explode.

R–410A recovery cylinders must be rated for 400 psig (use DOT 4BA400 or DOT 4BW400). Cylinders must be replaced or tested and date stamped every five years. Inspect cylinders for dents , rust, gouges or any visible damage. Do not use unsafe cylinders. Always transport cylinders secured in an upright position.

To prevent rusting, always store cylinders above the floor by using a platform or rack system. Pressurized cylinders should always be secured to prevent them from falling or rolling. The cylinders should be stored away from corrosive chemicals.

## Shipping

Before shipping any used refrigerant cylinders, check that the cylinder meets DOT standards, complete the shipping paperwork including the number of cylinders of each refrigerant, and properly label the cylinder with the type and amount of refrigerant. Cylinders should be transported in an upright position. Each cylinder must be marked with a DOT classification tag indicating it is a "2.2 non–flammable gas". Some states may require special shipping procedures to be followed based on their classification of used refrigerants. Check with the DOT in the state of origin.

## ASHRAE Standard 34

The American Society of Heating Refrigeration and Air Conditioning Engineers (ASHRAE) classifies refrigerants according to their toxicity and flammability.

Toxicity is based on the level to which an individual can be exposed over his or her working life without ill effects, defined as the Threshold Limit Value (TLV) and the Time–Weighted Average (TWA). Refrigerants for which toxicity has not been identified at concentrations at or below 400 ppm are classified Class A, while refrigerants showing evidence of toxicity at concentrations at or below 400 ppm are Class B. (Most of the refrigerants used throughout the industry are Class A.)

Flammability characteristics are divided into three numbered groups:

*Class 1*  refrigerants do not show flame propagation when tested in air at 14.7 psia (1 atmosphere) and 65° F.

*Class 2*  refrigerants have a lower flammability limit (LFL) of more than 0.00625 lb/ft$^3$ at 70° F and 14.7 psia, and the heat of combustion of less than 8,174 Btu/lb.

*Class 3*  refrigerants are highly flammable as defined by an LFL of less than or equal to 0.00625 lb/ft$^3$ at 70° F and 14.7 psia or a heat of combustion at or above 8,174 Btu/lb.

## Equipment Room / Job Site Safety

**ASHRAE Standard 15** requires the use of room sensors and alarms to detect refrigerant leaks. This standard includes all refrigerant safety groups.

Each machine room shall activate an alarm and mechanical ventilation before the refrigerant concentrations exceed the TLV and TWA.

Each refrigeration system must be protected with a safety relief device or some other means of safely relieving pressure. Multiple pressure relief valves are always installed in parallel, never in series. Pressure relief valves must always vent to the outdoors. ASHRAE 15 further defines five additional areas that should be covered.

- ◆ Monitors
- ◆ Alarms
- ◆ Ventilation
- ◆ Purge Venting
- ◆ Breathing Apparatus

*Monitors*

Each machinery room shall contain a detector, located in an area where refrigerant from a leak will concentrate, which will activate an alarm and cause mechanical ventilation to operate in accordance with 8.13.4 at a value not greater than the corresponding TLV– TWA (or toxicity measure consistent therewith).

*Alarms*

An alarm which activates at, or below, the AEL (Accepted Exposure Limit) for Group B1 refrigerants  When used, an oxygen alarm will be activated at not less than 19.5 percent (%) by volume.

*Ventilation*

Mechanical ventilation must be sized and used  per ASHRAE Standard 15R. This is typically not required for penthouse and lean–to applications.

*Purge  Venting*

Rupture disks and purges must be vented out of doors, using refrigerant compatible materials. A drip–leg and shutoff valve should be provided on vent piping.

*Breathing  Apparatus*

At least one approved self–contained breathing apparatus for emergency use should be at a convenient location within or very near an equipment room.

## Safety  Overview

Whenever possible, maintenance or cleaning of equipment should be performed without opening the sealed system. If a confined space must be entered, a fully qualified work team must be employed. The confined space entry form must be completed and posted at the work site. The following minimum guidelines MUST also be followed:

- ◆ Pull all fuses or safety jacks
- ◆ Lock and tag breakers and switches
- ◆ Test for toxic or flammable atmosphere
- ◆ Test for oxygen deficiency - 19.5 % minimum
- ◆ Assign and instruct observer
- ◆ Alert employees in the immediate area
- ◆ Provide fresh air supply
- ◆ Wear rescue harness
- ◆ Attach tie line
- ◆ Have rescue equipment nearby
- ◆ Wear protective clothing

To avoid pressure related incidents, be certain that the cylinders and system components carry the correct pressure rating for the refrigerant being used. Do not heat or store cylinders where they could reach temperatures over 125° F.

## R–410A Considerations

The introduction of R–410A as a replacement for R–22 has generated concerns as to its safety. R–410A has a much higher vapor pressure that most of the refrigerants in use today. The discharge pressure of R–410A is approximately 50% to 70% higher than R–22.

Along with the higher pressure rated system components we have covered in the preceding chapters, the storage cylinders (recovery and disposable) have a service rating of 400 psig. DOT 4BA 400 and DOT 4BW400 are acceptable cylinders.

The color code for R–410A disposable cylinders is ROSE. The color code for all recovery cylinders is GRAY with a YELLOW top. Tag all recovery cylinders with the type of refrigerant they contain.

The manifold gauge set used with R–410A requires a high side range of 800 psig and a low side range of 250 psig, with a 500 psig low side retard. Service hoses are required to have a pressure rating of 800 psig.

Avoid mixing R–410A with other refrigerants. Not only will the mixture be difficult or impossible to reclaim, the cumulative pressures may exceed the safety rating of the storage cylinder. To prevent cross contamination, use a self–purging recovery unit. Manifold gauges and hoses must be evacuated after each use.

Another option that will eliminate the chance of cross contamination is to dedicate service equipment for use on R–410A systems only.

## Material Safety Data Sheet (MSDS)

Material Safety Data Sheets are available for each refrigerant from the manufacturer.

The service technician should be familiar with the hazards of working with a given refrigerant as indicated on the MSDS.

## MSDS Overview

### Toxicity

The Program for Alternative Fluorocarbon Toxicity Testing (PAFT) is an international consortium of refrigerant producers. The data developed by PAFT's III and V have confirmed the toxicity of R–410A to be in the low toxicity range.

### Flammability

The ASHRAE Safety Group Classification (ASHRAE Standard 34) for R–410A is A1/A1. Underwriters Laboratory (UL) lists R–410A as "Practically Non Flammable" and Department of Transportation (DOT) considers R–410A non flammable and tanks carry a green label.

### Combustibility

Although R–410A is not flammable, at pressures above 1 atmosphere, mixtures of R–410A and air can become combustible.

Never leak check with a mixture of R–410A and air.

### Ingestion

It is very unlikely that ingestion of R–410A or any refrigerant would occur because of their physical properties. If ingestion occurs, do not induce vomiting. Seek medical attention immediately.

### Skin or Eye Contact

As with all refrigerants, care should be taken to avoid liquid contact with skin or eyes. Frost–bite could occur if the liquid experiences direct expansion. Promptly flush eyes or skin with lukewarm water. Seek medical attention. *(POE oils can cause skin irritations. It is recommended to use appropriate gloves and safety glasses when handling POE lubricants. When finished handling the POE lubricant, thoroughly wash with soap and water to remove any residual oil.)*

### Inhalation

Inhaling high concentrations of refrigerant vapors initially attacks the central nervous system, creating a narcotic effect. A feeling of intoxication and dizziness with loss of coordination and slurred speech are symptoms. Cardiac irregularities, unconsciousness and ultimate death can result from the breathing of this concentration. If any of these symptoms become evident, move to fresh air and seek medical help immediately.

### Refrigerant Decomposition

When refrigerants are exposed to high temperatures from open flames or resistive heater elements, decomposition occurs. Decomposition produces toxic and irritating compounds, such as hydrogen chloride (with chlorinated refrigerants such as CFCs & HCFCs) and hydrogen fluoride (with CFCs, HCFCs & HFCs). The acidic vapors produced are dangerous and the area should be evacuated and ventilated to prevent exposure to personnel.

## Environmental Considerations

Treatment or disposal of wastes generated by the use of R–410A may require special consideration. For specific information, refer to the Material Safety Data Sheet (MSDS). If discarded unused, R–410A is not considered a hazardous waste by the Resource Conservation and Recovery Act (RCRA). However, the disposal of R–410A may be subject to federal, state and local regulations. Appropriate regulatory agencies should be consulted before disposing of waste materials.

# Honeywell

## Gentron®
## AZ-20 (R-410A)

## Material Safety Data Sheet
## (MSDS)

# *Material Safety Data Sheet*

## Genetron® AZ-20 (R-410A)

### 1. CHEMICAL PRODUCT AND COMPANY IDENTIFICATION

**PRODUCT NAME:** Genetron® AZ-20 (R-410A)
**OTHER/GENERIC NAMES:** R-410A
**PRODUCT USE:** Refrigerant
**MANUFACTURER:** Honeywell
101 Columbia Road
Box 1053
Morristown, New Jersey 07962-1053

**FOR MORE INFORMATION CALL:**
(Monday-Friday, 9:00am-5:00pm)
Product Safety Department
1-800-707-4555

**IN CASE OF EMERGENCY CALL:**
(24 Hours/Day, 7 Days/Week)
973-455-2000

### 2. COMPOSITION/INFORMATION ON INGREDIENTS

| INGREDIENT NAME | CAS NUMBER | WEIGHT % |
|---|---|---|
| Difluoromethane | 75-10-5 | 50 |
| Pentafluoroethane | 354-33-6 | 50 |

Trace impurities and additional material names not listed above may also appear in Section 15 toward the end of the MSDS. These materials may be listed for local "Right-To-Know" compliance and for other reasons.

### 3. HAZARDS IDENTIFICATION

**EMERGENCY OVERVIEW: Colorless, volatile liquid with ethereal and faint sweetish odor. Non-flammable material. Overexposure may cause dizziness and loss of concentration. At higher levels, CNS depression and cardiac arrhythmia may result from exposure. Vapors displace air and can cause asphyxiation in confined spaces. At higher temperatures, (>250°C), decomposition products may include Hydrofluoric Acid (HF) and carbonyl halides**

#### POTENTIAL HEALTH HAZARDS

**SKIN:** Irritation would result from a defatting action on tissue. Liquid contact could cause frostbite.

**EYES:** Liquid contact can cause severe irritation and frostbite. Mist may irritate.

**INHALATION:** Genetron AZ-20 (R-410A) is low in acute toxicity in animals. When oxygen levels in air are reduced to 12–14% by displacement, symptoms of asphyxiation, loss of coordination, increased pulse rate and deeper respiration will occur. At high levels, cardiac arrhythmia may occur.

**INGESTION:**   Ingestion is unlikely because of the low boiling point of the material. Should it occur, discomfort in the gastrointestinal tract from rapid evaporation of the material and consequent evolution of gas would result. Some effects of inhalation and skin exposure would be expected.

**DELAYED EFFECTS:**   None known

**Ingredients found on one of the OSHA designated carcinogen lists are listed below.**

| INGREDIENT NAME | NTP STATUS | IARC STATUS | OSHA LIST |
|---|---|---|---|
| No ingredients listed in this section | | | |

## 4.   FIRST AID MEASURES

**SKIN:**   Promptly flush skin with water until all chemical is removed. If there is evidence of frostbite, bathe (do not rub) with lukewarm (not hot) water. If water is not available, cover with a clean, soft cloth or similar covering. Get medical attention if symptoms persist.

**EYES:**   Immediately flush eyes with large amounts of water for at least 15 minutes (in case of frostbite water should be lukewarm, not hot) lifting eyelids occasionally to facilitate irrigation. Get medical attention if symptoms persist.

**INHALATION:**   Immediately remove to fresh air. If breathing has stopped, give artificial respiration. Use oxygen as required, provided a qualified operator is available. Get medical attention. Do not give epinephrine (adrenaline).

**INGESTION:**   Ingestion is unlikely because of the physical properties and is not expected to be hazardous. Do not induce vomiting unless instructed to do so by a physician.

**ADVICE TO PHYSICIAN:**   Because of the possible disturbances of cardiac rhythm, catecholamine drugs, such as epinephrine, should be used with special caution and only in situations of emergency life support. Treatment of overexposure should be directed at the control of symptoms and the clinical conditions.

## 5.   FIRE FIGHTING MEASURES

### FLAMMABLE PROPERTIES

**FLASH POINT:**   Gas, not applicable per DOT regulations
**FLASH POINT METHOD:**   Not applicable
**AUTOIGNITION TEMPERATURE:**   >750°C
**UPPER FLAME LIMIT (volume % in air):**   None by ASTM D-56-82
**LOWER FLAME LIMIT (volume % in air):**   None by ASTM E-681
**FLAME PROPAGATION RATE (solids):**   Not applicable
**OSHA FLAMMABILITY CLASS:**   Not applicable

### EXTINGUISHING MEDIA:
Use any standard agent – choose the one most appropriate for type of surrounding fire (material itself is not flammable)

# MATERIAL SAFETY DATA SHEET
# Genetron® AZ-20 (R-410A)

**UNUSUAL FIRE AND EXPLOSION HAZARDS:**

Genetron AZ-20 (R-410A) is not flammable at ambient temperatures and atmospheric pressure. However, this material will become combustible when mixed with air under pressure and exposed to strong ignition sources.

Contact with certain reactive metals may result in formation of explosive or exothermic reactions under specific conditions (e.g. very high temperatures and/or appropriate pressures).

**SPECIAL FIRE FIGHTING PRECAUTIONS/INSTRUCTIONS:**

Firefighters should wear self-contained, NIOSH-approved breathing apparatus for protection against possible toxic decomposition products. Proper eye and skin protection should be provided. Use water spray to keep fire-exposed containers cool.

## 6. ACCIDENTAL RELEASE MEASURES

**IN CASE OF SPILL OR OTHER RELEASE:** (Always wear recommended personal protective equipment.)

Evacuate unprotected personnel. Protected personnel should remove ignition sources and shut off leak, if without risk, and provide ventilation. Unprotected personnel should not return until air has been tested and determined safe, including low-lying areas.

**Spills and releases may have to be reported to Federal and/or local authorities. See Section 15 regarding reporting requirements.**

## 7. HANDLING AND STORAGE

**NORMAL HANDLING:** (Always wear recommended personal protective equipment.)

Avoid breathing vapors and liquid contact with eyes, skin or clothing. Do not puncture or drop cylinders, expose them to open flame or excessive heat. Use authorized cylinders only. Follow standard safety precautions for handling and use of compressed gas cylinders.

Genetron AZ-20 (R-410A) should not be mixed with air above atmospheric pressure for leak testing or any other purpose.

**STORAGE RECOMMENDATIONS:**

Store in a cool, well-ventilated area of low fire risk and out of direct sunlight. Protect cylinder and its fittings from physical damage. Storage in subsurface locations should be avoided. Close valve tightly after use and when empty.

## 8. EXPOSURE CONTROLS/PERSONAL PROTECTION

**ENGINEERING CONTROLS:**

Provide local ventilation at filling zones and areas where leakage is probable. Mechanical (general) ventilation may be adequate for other operating and storage areas.

**PERSONAL PROTECTIVE EQUIPMENT**

**SKIN PROTECTION:**

Skin contact with refrigerant may cause frostbite. General work clothing and gloves (leather) should provide adequate protection. If prolonged contact with the liquid or gas is anticipated, insulated gloves constructed of PVA, neoprene or butyl rubber should be used. Any contaminated clothing should be promptly removed and washed before reuse.

**Honeywell**

## MATERIAL SAFETY DATA SHEET
# Genetron® AZ-20 (R-410A)

**EYE PROTECTION:**
For normal conditions, wear safety glasses. Where there is reasonable probability of liquid contact, wear chemical safety goggles.

**RESPIRATORY PROTECTION:**
None generally required for adequately ventilated work situations. For accidental release or non-ventilated situations, or release into confined space, where the concentration may be above the PEL of 1,000 ppm, use a self-contained, NIOSH - approved breathing apparatus or supplied air respirator. For escape: use the former or a NIOSH-approved gas mask with organic vapor canister.

**ADDITIONAL RECOMMENDATIONS:**
Where contact with liquid is likely, such as in a spill or leak, impervious boots and clothing should be worn. High dose-level warning signs are recommended for areas of principle exposure. Provide eyewash stations and quick-drench shower facilities at convenient locations. For tank cleaning operations, see OSHA regulations, 29 CFR 1910.132 and 29 CFR 1910.133.

**EXPOSURE GUIDELINES**

| INGREDIENT NAME | ACGIH TLV | OSHA PEL | OTHER LIMIT |
|---|---|---|---|
| Difluoromethane | None | None | *1000 ppm TWA |
| Pentafluoroethane | None | None | (8hr) |
| | | | **1000 ppm TWA |
| | | | (8hr) |

\* = Limit established by AlliedSignal.
\*\* = Workplace Environmental Exposure Level (AIHA).
\*\*\* = Biological Exposure Index (ACGIH).

**OTHER EXPOSURE LIMITS FOR POTENTIAL DECOMPOSITION PRODUCTS:**
Hydrogen Fluoride: ACGIH TLV: 3 ppm ceiling

## 9. PHYSICAL AND CHEMICAL PROPERTIES

**APPEARANCE:** Clear, colorless liquid and vapor
**PHYSICAL STATE:** Gas at ambient temperatures
**MOLECULAR WEIGHT:** 72.6
**CHEMICAL FORMULA:** $CH_2F_2$
$CHF_2CF_3$
**ODOR:** Faint ethereal odor
**SPECIFIC GRAVITY (water = 1.0):** 1.08 @ 21.1°C (70°F)
**SOLUBILITY IN WATER (weight %):** Unknown
**pH:** Neutral
**BOILING POINT:** -48.5°C (-55.4°F)
**FREEZING POINT:** Not Determined
**VAPOR PRESSURE:** 215.3 psia @ 70°F
490.2 psia @ 130°F
**VAPOR DENSITY (air = 1.0):** 3.0
**EVAPORATION RATE:** >1 **COMPARED TO:** $CCl_4 = 1$

## MATERIAL SAFETY DATA SHEET
## Genetron® AZ-20 (R-410A)

---

**% VOLATILES:**     100
**FLASH POINT:**     Not applicable
    (Flash point method and additional flammability data are found in Section 5.)

---

### 10. STABILITY AND REACTIVITY

**NORMALLY STABLE? (CONDITIONS TO AVOID):**
    The product is stable.
    Do not mix with oxygen or air above atmospheric pressure. Any source of high temperature, such as lighted cigarettes, flames, hot spots or welding may yield toxic and/or corrosive decomposition products.

**INCOMPATIBILITIES:**
    (Under specific conditions: e.g. very high temperatures and/or appropriate pressures) – Freshly abraded aluminum surfaces (may cause strong exothermic reaction). Chemically active metals: potassium, calcium, powdered aluminum, magnesium and zinc.

**HAZARDOUS DECOMPOSITION PRODUCTS:**
    Halogens, halogen acids and possibly carbonyl halides.

**HAZARDOUS POLYMERIZATION:**
    Will not occur.

---

### 11. TOXICOLOGICAL INFORMATION

**IMMEDIATE (ACUTE) EFFECTS:**
    $LC_{50}$ : 4 hr. (rat) - ≥520,000 ppm (difluoromethane)
    Cardiac Sensitization threshold (dog) ≥100,000 ppm (pentafluoroethane)

**DELAYED (SUBCHRONIC AND CHRONIC) EFFECTS:**
    Teratology - negative
    Subchronic inhalation (rat) NOEL - 50,000 ppm

**OTHER DATA:**
    Not active in four genetic studies

---

### 12. ECOLOGICAL INFORMATION

**Degradability (BOD):** Genetron AZ-20 (R-410A) is a gas at room temperature; therefore, it is unlikely to remain in water.
**Octanol Water Partition Coefficient:** Log $P_{ow}$ = 1.48 (pentafluoroethane), 0.21 (difluoromethane)

### 13. DISPOSAL CONSIDERATIONS

**RCRA**

    **Is the unused product a RCRA hazardous waste if discarded?**    Not a hazardous waste
    **If yes, the RCRA ID number is:**    Not applicable

---

## MATERIAL SAFETY DATA SHEET
# Genetron® AZ-20 (R-410A)

**OTHER DISPOSAL CONSIDERATIONS:**
Disposal must comply with federal, state, and local disposal or discharge laws. Genetron AZ-20 (R-410A) is subject to U.S. Environmental Protection Agency Clean Air Act Regulations Section 608 in 40 CFR Part 82 regarding refrigerant recycling.

The information offered here is for the product as shipped. Use and/or alterations to the product such as mixing with other materials may significantly change the characteristics of the material and alter the RCRA classification and the proper disposal method.

## 14. TRANSPORT INFORMATION

**US DOT HAZARD CLASS:** US DOT PROPER SHIPPING NAME: Liquified gas, n.o.s. (Pentafluoroethane, Difluoromethane)
US DOT HAZARD CLASS: 2.2
US DOT PACKING GROUP: Not applicable

**US DOT ID NUMBER:** UN3163

For additional information on shipping regulations affecting this material, contact the information number found in Section 1.

## 15. REGULATORY INFORMATION

### TOXIC SUBSTANCES CONTROL ACT (TSCA)

**TSCA INVENTORY STATUS:** Components listed on the TSCA inventory

**OTHER TSCA ISSUES:** None

### SARA TITLE III/CERCLA

"Reportable Quantities" (RQs) and/or "Threshold Planning Quantities" (TPQs) exist for the following ingredients.

| INGREDIENT NAME | SARA/CERCLA RQ (lb.) | SARA EHS TPQ (lb.) |
|---|---|---|
| No ingredients listed in this section | | |

**Spills or releases resulting in the loss of any ingredient at or above its RQ requires immediate notification to the National Response Center [(800) 424-8802] and to your Local Emergency Planning Committee.**

**SECTION 311 HAZARD CLASS:** IMMEDIATE
PRESSURE

**SARA 313 TOXIC CHEMICALS:**
The following ingredients are SARA 313 "Toxic Chemicals". CAS numbers and weight percents are found in Section 2.

| INGREDIENT NAME | COMMENT |
|---|---|
| No ingredients listed in this section | |

# MATERIAL SAFETY DATA SHEET
## Genetron® AZ-20 (R-410A)

**STATE RIGHT-TO-KNOW**

In addition to the ingredients found in Section 2, the following are listed for state right-to-know purposes.

| INGREDIENT NAME | WEIGHT % | COMMENT |
|---|---|---|
| No ingredients listed in this section | | |

**ADDITIONAL REGULATORY INFORMATION:**

Genetron AZ-20 (R-410A) is subject to U.S. Environmental Protection Agency Clean Air Act Regulations at 40 CFR Part 82.

**WARNING:** Contains pentafluoroethane (HFC-125) and difluoromethane (HFC-32), greenhouse gases which may contribute to global warming
**Do Not vent** to the atmosphere. To comply with provisions of the U.S. Clean Air Act, any residual must be recovered.

**WHMIS CLASSIFICATION (CANADA):**

This product has been evaluated in accordance with the hazard criteria of the CPR and the MSDS contains all the information required by the CPR.

**FOREIGN INVENTORY STATUS:**

EU – EINECS # 2065578 (HFC-125)

## 16. OTHER INFORMATION

**CURRENT ISSUE DATE:**   January, 2000
**PREVIOUS ISSUE DATE:**   February, 1999
**CHANGES TO MSDS FROM PREVIOUS ISSUE DATE ARE DUE TO THE FOLLOWING:**
   Section 1:  New company name
   Section 16:  Modified NFPA and HMIS codes


**OTHER INFORMATION:**     HMIS Classification:  Health – 1, Flammability – 1, Reactivity – 0
     NFPA Classification:  Health – 2, Flammability – 1, Reactivity – 0
     ANSI/ASHRAE 34 Safety Group – A1

     Regulatory Standards:
     1.   OSHA regulations for compressed gases:  29 CFR 1910.101
     2.   DOT classification per 49 CFR 172.101

     Toxicity information per PAFT Testing

---

# Honeywell

## Gentron®
## 407C (R-407C)

## Material Safety Data Sheet
## (MSDS)

# *Material Safety Data Sheet*

## Genetron® 407C

### 1. CHEMICAL PRODUCT AND COMPANY IDENTIFICATION

**PRODUCT NAME:**  Genetron® 407C
**OTHER/GENERIC NAMES:**  R-407C
**PRODUCT USE:**  Refrigerant
**MANUFACTURER:**  Honeywell
101 Columbia Road
Box 1053
Morristown, New Jersey  07962-1053

**FOR MORE INFORMATION CALL:**
(Monday-Friday, 9:00am-5:00pm)
Product Safety Department
1-800-707-4555

**IN CASE OF EMERGENCY CALL:**
(24 Hours/Day, 7 Days/Week)
1-800-707-4555
CHEMTREC: 1-800-424-9300

### 2. COMPOSITION/INFORMATION ON INGREDIENTS

| INGREDIENT NAME | CAS NUMBER | WEIGHT % |
|---|---|---|
| Difluoromethane (HFC-32) | 75-10-5 | 23 |
| Pentafluoroethane (HFC-125) | 354-33-6 | 23 |
| 1,1,1,2-Tetrafluoroethane (HFC-134a) | 811-97-2 | 52 |

Trace impurities and additional material names not listed above may also appear in Section 15 toward the end of the MSDS. These materials may be listed for local "Right-To-Know" compliance and for other reasons.

### 3. HAZARDS IDENTIFICATION

**EMERGENCY OVERVIEW:  Colorless, volatile liquid with ethereal and faint sweetish odor.  Non-flammable material.  Overexposure may cause dizziness and loss of concentration.  At higher levels, CNS depression and cardiac arrhythmia may result from exposure.  Vapors displace air and can cause asphyxiation in confined spaces.  At higher temperatures, (>250°C), decomposition products may include Hydrofluoric Acid (HF) and carbonyl halides**

<u>POTENTIAL HEALTH HAZARDS</u>

**SKIN:**  Irritation would result from a defatting action on tissue.  Liquid contact could cause frostbite.

**EYES:**  Liquid contact can cause severe irritation and frostbite.  Mist may irritate.

**INHALATION:**  Genetron 407C is low in acute toxicity in animals.  When oxygen levels in air are reduced to 12–14% by displacement, symptoms of asphyxiation, loss of coordination, increased pulse rate and deeper respiration will occur.  At high levels, cardiac arrhythmia may occur.

---

MSDS Number:  GTRN-0022
Current Issue Date: January 2000

# MATERIAL SAFETY DATA SHEET
# Genetron® 407C

**INGESTION:** Ingestion is unlikely because of the low boiling point of the material. Should it occur, discomfort in the gastrointestinal tract from rapid evaporation of the material and consequent evolution of gas would result. Some effects of inhalation and skin exposure would be expected.

**DELAYED EFFECTS:** None known

**Ingredients found on one of the OSHA designated carcinogen lists are listed below.**

| INGREDIENT NAME | NTP STATUS | IARC STATUS | OSHA LIST |
|---|---|---|---|
| No ingredients listed in this section | | | |

## 4. FIRST AID MEASURES

**SKIN:** Promptly flush skin with water until all chemical is removed. If there is evidence of frostbite, bathe (do not rub) with lukewarm (not hot) water. If water is not available, cover with a clean, soft cloth or similar covering. Get medical attention if symptoms persist.

**EYES:** Immediately flush eyes with large amounts of water for at least 15 minutes (in case of frostbite water should be lukewarm, not hot) lifting eyelids occasionally to facilitate irrigation. Get medical attention if symptoms persist.

**INHALATION:** Immediately remove to fresh air. If breathing has stopped, give artificial respiration. Use oxygen as required, provided a qualified operator is available. Get medical attention. Do not give epinephrine (adrenaline).

**INGESTION:** Ingestion is unlikely because of the physical properties and is not expected to be hazardous. Do not induce vomiting unless instructed to do so by a physician.

**ADVICE TO PHYSICIAN:** Because of the possible disturbances of cardiac rhythm, catecholamine drugs, such as epinephrine, should be used with special caution and only in situations of emergency life support. Treatment of overexposure should be directed at the control of symptoms and the clinical conditions.

## 5. FIRE FIGHTING MEASURES

### FLAMMABLE PROPERTIES

**FLASH POINT:** Gas, not applicable per DOT regulations
**FLASH POINT METHOD:** Not applicable
**AUTOIGNITION TEMPERATURE:** Unknown for mixture
**UPPER FLAME LIMIT (volume % in air):** None*
**LOWER FLAME LIMIT (volume % in air):** None*
　　　　　　　　　　　　　　　　　　*Based on ASHRAE Standard 34 with match ignition
**FLAME PROPAGATION RATE (solids):** Not applicable
**OSHA FLAMMABILITY CLASS:** Not applicable

### EXTINGUISHING MEDIA:
Use any standard agent – choose the one most appropriate for type of surrounding fire (material itself is not flammable)

## MATERIAL SAFETY DATA SHEET
# Genetron® 407C

**UNUSUAL FIRE AND EXPLOSION HAZARDS:**
Genetron 407C is not flammable at ambient temperatures and atmospheric pressure. However, this material will become combustible when mixed with air under pressure and exposed to strong ignition sources.
Contact with certain reactive metals may result in formation of explosive or exothermic reactions under specific conditions (e.g. very high temperatures and/or appropriate pressures).

**SPECIAL FIRE FIGHTING PRECAUTIONS/INSTRUCTIONS:**
Firefighters should wear self-contained, NIOSH-approved breathing apparatus for protection against possible toxic decomposition products. Proper eye and skin protection should be provided. Use water spray to keep fire-exposed containers cool.

## 6. ACCIDENTAL RELEASE MEASURES

**IN CASE OF SPILL OR OTHER RELEASE:**  (Always wear recommended personal protective equipment.)
Evacuate unprotected personnel. Protected personnel should remove ignition sources and shut off leak, if without risk, and provide ventilation. Unprotected personnel should not return until air has been tested and determined safe, including low-lying areas.

**Spills and releases may have to be reported to Federal and/or local authorities. See Section 15 regarding reporting requirements.**

## 7. HANDLING AND STORAGE

**NORMAL HANDLING:**  (Always wear recommended personal protective equipment.)
Avoid breathing vapors and liquid contact with eyes, skin or clothing. Do not puncture or drop cylinders, expose them to open flame or excessive heat. Use authorized cylinders only. Follow standard safety precautions for handling and use of compressed gas cylinders.

Genetron 407C should not be mixed with air above atmospheric pressure for leak testing or any other purpose.

**STORAGE RECOMMENDATIONS:**
Store in a cool, well-ventilated area of low fire risk and out of direct sunlight. Protect cylinder and its fittings from physical damage. Storage in subsurface locations should be avoided. Close valve tightly after use and when empty.

## 8. EXPOSURE CONTROLS/PERSONAL PROTECTION

**ENGINEERING CONTROLS:**
Provide local ventilation at filling zones and areas where leakage is probable. Mechanical (general) ventilation may be adequate for other operating and storage areas.

**PERSONAL PROTECTIVE EQUIPMENT**

**SKIN PROTECTION:**
Skin contact with refrigerant may cause frostbite. General work clothing and gloves (leather) should provide adequate protection. If prolonged contact with the liquid or gas is anticipated, insulated gloves constructed of PVA, neoprene or butyl rubber should be used. Any contaminated clothing should be promptly removed and washed before reuse.

## MATERIAL SAFETY DATA SHEET
## Genetron® 407C

**EYE PROTECTION:**

For normal conditions, wear safety glasses. Where there is reasonable probability of liquid contact, wear chemical safety goggles.

**RESPIRATORY PROTECTION:**

None generally required for adequately ventilated work situations. For accidental release or non-ventilated situations, or release into confined space, where the concentration may be above the PEL of 1,000 ppm, use a self-contained, NIOSH - approved breathing apparatus or supplied air respirator. For escape: use the former or a NIOSH-approved gas mask with organic vapor canister.

**ADDITIONAL RECOMMENDATIONS:**

Where contact with liquid is likely, such as in a spill or leak, impervious boots and clothing should be worn. High dose-level warning signs are recommended for areas of principle exposure. Provide eyewash stations and quick-drench shower facilities at convenient locations. For tank cleaning operations, see OSHA regulations, 29 CFR 1910.132 and 29 CFR 1910.133.

**EXPOSURE GUIDELINES**

| INGREDIENT NAME | ACGIH TLV | OSHA PEL | OTHER LIMIT |
|---|---|---|---|
| Difluoromethane | None | None | **1000 ppm TWA (8hr) |
| Pentafluoroethane | None | None | **1000 ppm TWA (8hr) |
| 1,1,1,2-Tetrafluoroethane | None | None | **1000 ppm TWA (8hr) |

\*     = Limit established by Honeywell.
\*\*    = Workplace Environmental Exposure Level (AIHA).
\*\*\* = Biological Exposure Index (ACGIH).

**OTHER EXPOSURE LIMITS FOR POTENTIAL DECOMPOSITION PRODUCTS:**

Hydrogen Fluoride: ACGIH TLV: 3 ppm ceiling

## 9.   PHYSICAL AND CHEMICAL PROPERTIES

**APPEARANCE:** Clear, colorless liquid and vapor
**PHYSICAL STATE:**         Gas at ambient temperatures
**MOLECULAR WEIGHT:**    86.2
**CHEMICAL FORMULA:**    $CH_2F_2$, $CF_3CHF_2$, $CH_2FCF_3$
**ODOR:**                 Faint ethereal odor
**SPECIFIC GRAVITY (water = 1.0):**       1.16 @ 21.1°C (70°F)
**SOLUBILITY IN WATER (weight %):**       Unknown
**pH:**   Neutral
**BOILING POINT:**              -43°C (-45.4°F)
**FREEZING POINT:**            Not Determined
**VAPOR PRESSURE:**          156.2 psia @ 70°F
                                356.7 psia @ 130°F
**VAPOR DENSITY (air = 1.0):**       3.0

## MATERIAL SAFETY DATA SHEET
## Genetron® 407C

---

**EVAPORATION RATE:**          >1          **COMPARED TO:**    $CCl_4 = 1$
**% VOLATILES:**                  100
**FLASH POINT:**                Not applicable
    (Flash point method and additional flammability data are found in Section 5.)

---

### 10. STABILITY AND REACTIVITY

**NORMALLY STABLE? (CONDITIONS TO AVOID):**
    The product is stable.
    Do not mix with oxygen or air above atmospheric pressure. Any source of high temperature, such as lighted cigarettes, flames, hot spots or welding may yield toxic and/or corrosive decomposition products.

**INCOMPATIBILITIES:**
    (Under specific conditions: e.g. very high temperatures and/or appropriate pressures) – Freshly abraded aluminum surfaces (may cause strong exothermic reaction). Chemically active metals: potassium, calcium, powdered aluminum, magnesium and zinc.

**HAZARDOUS DECOMPOSITION PRODUCTS:**
    Halogens, halogen acids and possibly carbonyl halides.

**HAZARDOUS POLYMERIZATION:**
    Will not occur.

---

### 11. TOXICOLOGICAL INFORMATION

**IMMEDIATE (ACUTE) EFFECTS:**
    HFC-32:     $LC_{50}$ : 4 hr. (rat) - 520,000 ppm
                    Cardiac Sensitization threshold (dog) 350,000 ppm.
    HFC-125:    $LC_{50}$ : 4 hr. (rat) - > 800,000 ppm
                    Cardiac Sensitization threshold (dog) 75,000 ppm.
    HFC-134a:  $LC_{50}$ : 4 hr. (rat) - > 500,000 ppm
                    Cardiac Sensitization threshold (dog) > 80,000 ppm.

**DELAYED (SUBCHRONIC AND CHRONIC) EFFECTS:**
    HFC-32:     Teratogenic NOEL (rat and rabbit) - 50,000 ppm
                   Subchronic inhalation (rat) NOEL - 50,000 ppm
    HFC-125:    Teratogenic NOEL (rat and rabbit) - 50,000 ppm
                   Subchronic inhalation (rat) NOEL - ≥50,000 ppm
                   Chronic NOEL – 10,000 ppm

    HFC-134a:  Teratogenic NOEL (rat and rabbit) – 40,000 ppm
                   Subchronic inhalation (rat) NOEL - 50,000 ppm
                   Chronic NOEL – 10,000 ppm

**OTHER DATA:**
    HFC-32, HFC-125, HFC-134a: Not active in four genetic studies

---

# MATERIAL SAFETY DATA SHEET
## Genetron® 407C

## 12. ECOLOGICAL INFORMATION

**Degradability (BOD):** Genetron 407C is a gas at room temperature; therefore, it is unlikely to remain in water.
**Octanol Water Partition Coefficient:** Unknown for mixture

## 13. DISPOSAL CONSIDERATIONS

### RCRA

**Is the unused product a RCRA hazardous waste if discarded?**     Not a hazardous waste
**If yes, the RCRA ID number is:**     Not applicable

### OTHER DISPOSAL CONSIDERATIONS:
Disposal must comply with federal, state, and local disposal or discharge laws. Genetron 407C is subject to U.S. Environmental Protection Agency Clean Air Act Regulations Section 608 in 40 CFR Part 82 regarding refrigerant recycling.

The information offered here is for the product as shipped. Use and/or alterations to the product such as mixing with other materials may significantly change the characteristics of the material and alter the RCRA classification and the proper disposal method.

## 14. TRANSPORT INFORMATION

**US DOT PROPER SHIPPING NAME:**     Refrigerant gas R 407C
**US DOT HAZARD CLASS:**     2.2
**US DOT PACKING GROUP:**     Not applicable
**US DOT ID NUMBER:**     UN3340

For additional information on shipping regulations affecting this material, contact the information number found in Section 1.

## 15. REGULATORY INFORMATION

### TOXIC SUBSTANCES CONTROL ACT (TSCA)

**TSCA INVENTORY STATUS:**     Components listed on the TSCA inventory

**OTHER TSCA ISSUES:**     None

### SARA TITLE III/CERCLA

"Reportable Quantities" (RQs) and/or "Threshold Planning Quantities" (TPQs) exist for the following ingredients.

| INGREDIENT NAME | SARA/CERCLA RQ (lb.) | SARA EHS TPQ (lb.) |
|---|---|---|
| No ingredients listed in this section | | |

**Spills or releases resulting in the loss of any ingredient at or above its RQ requires immediate notification to the National Response Center [(800) 424-8802] and to your Local Emergency Planning Committee.**

## MATERIAL SAFETY DATA SHEET
# Genetron® 407C

**SECTION 311 HAZARD CLASS:**     IMMEDIATE
                                                      PRESSURE

**SARA 313 TOXIC CHEMICALS**:
The following ingredients are SARA 313 "Toxic Chemicals".  CAS numbers and weight percents are found in Section 2.

| INGREDIENT NAME | COMMENT |
|---|---|
| No ingredients listed in this section | |

### STATE RIGHT-TO-KNOW

In addition to the ingredients found in Section 2, the following are listed for state right-to-know purposes.

| INGREDIENT NAME | WEIGHT % | COMMENT |
|---|---|---|
| No ingredients listed in this section | | |

**ADDITIONAL REGULATORY INFORMATION:**
Genetron 407C is subject to U.S. Environmental Protection Agency Clean Air Act Regulations at 40 CFR Part 82.

**WARNING:**  Contains pentafluoroethane (HFC-125), 1,1,1-trifluoroethane, tetrafluoroethane,  greenhouse gases which may contribute to global warming
**Do Not vent** to the atmosphere.  To comply with provisions of the U.S. Clean Air Act, any residual must be recovered.

**WHMIS CLASSIFICATION (CANADA):**
This product has been evaluated in accordance with the hazard criteria of the CPR and the MSDS contains all the information required by the CPR.

**FOREIGN INVENTORY STATUS:**
EU – EINECS # 2065578 – HFC-125
              # 2008394 – HFC-32
              # 223770 – HFC134a

## 16.  OTHER INFORMATION

**CURRENT ISSUE DATE:**     January, 2000
**PREVIOUS ISSUE DATE:**     August, 1999

**CHANGES TO MSDS FROM PREVIOUS ISSUE DATE ARE DUE TO THE FOLLOWING:**
Section 1:  New company name
Section 14:  New hazardous material shipping description
Section 16:  Modified NFPA and HMIS codes

# MATERIAL SAFETY DATA SHEET
## Genetron® 407C

**OTHER INFORMATION:**

HMIS Classification:  Health – 1, Flammability – 1, Reactivity – 0
NFPA Classification:  Health – 2, Flammability – 1, Reactivity – 0
ANSI/ASHRAE 34 Safety Group – A1

<u>Regulatory Standards</u>:
1.  OSHA regulations for compressed gases:  29 CFR 1910.101
2.  DOT classification per 49 CFR 172.101

Toxicity information per PAFT Testing

**6**

# Appendix I

## History of Refrigerants

# History of Refrigerants

Even though water and ice were the first refrigerants, ether was the first commercial refrigerant. In 1850, ice was made by evaporating ether under a vacuum produced by a steam driven pump. By 1855, there was an ether machine that could produce a maximum of 2,000 pounds of ice per day. This was a vapor compression process that used volatile etheral fluid as the refrigerant. The refrigerant was then condensed and reused generating no wasted ether. Many other etheral based machines were developed including one that transported chilled meats across the sea from France to South America by ship. Because ether operated in a vacuum and was extremely flammable, the last ether machine was made in 1902.

The following decades saw numerous compounds tested as refrigerants. In fact, one can safely say that just about every volatile fluid was tested as a refrigerant. Mechanical vapor compression refrigeration was firmly established by the turn of the century. However, these early refrigerants all had disadvantages and advantages. Listed below are some of the most popular refrigerants used according to date.

### 1850–1856
An ammonia (R–717) machine received a patent. Good thermodynamic properties and low costs made ammonia a widely used refrigerant even today. Ammonia is very irritable to the mucous membranes and is flammable in certain concentrations.

### 1882–1886
Carbon dioxide (R–744) was used on British ships to the 1940s until it was replaced by chlorofluorocarbons. Carbon dioxide systems never saw widespread usage in the United States.

### 1880–1940
Sulfur dioxide (R–764) has an advantage of low costs and low operating pressures for warmer climates, but high enough pressures to remain out of a vacuum. Sulfur dioxide was used as a refrigerant in household refrigerators around 1900. One drawback to sulfur dioxide systems was that the refrigerant reacts with moisture and forms sulfurous acid. This resulted in a lot of seized compressors. Even though sulfur dioxide is a toxic refrigerant, the smallest of leaks can be detected by smell. This refrigerant remained popular in smaller units until the 1940s.

**1890**

Methyl chloride (R–40) was used sparingly in the United States until 1910. Its earlier use was in shipping meats across the sea. Methyl chloride has a sweet, etheral odor and has somewhat of an anesthetic effect when inhaled. It is also mildly flammable. Leaks in larger systems had many fatal results. This refrigerant also reacted with the aluminum in hermetic motors in the 1940s. Its use declined in the late 1930s.

**1880–1890**

Ethyl chloride (R–160) was used as an anesthetic. Liquid refrigerant was sprayed on the skin before surgery. This refrigerant was used in some household refrigerators after 1900.

Some other refrigerants that were experimented with and often used for short periods of time include:

| | |
|---|---|
| **methylamine** | **naphtha** |
| **nitrous oxide** | **methyl acetate** |
| **butane** | **pentane** |
| **propylene** | **isobutylene** |
| **carbon tetrachloride** | **gasoline** |
| **dielene** | **trielene** |
| **ethyl bromide** | |

Most of the refrigerants mentioned so far are either toxic, flammable, or smelled horrible. There was always some health risk when incorporating these refrigerants in the home. The refrigeration industry needed newer, safer refrigerants. The Frigidaire Company asked General Research Laboratories to develop a safe refrigerant. This was the beginning of the chlorofluorocarbons (CFC) refrigerants. The research team, headed by Thomas Midgley of General Motors, settled on R–12 as the refrigerant most suitable for commercial use. The first use of R–12 was in small ice cream applications in 1931. R–12 soon became a commercial refrigerant for room coolers. In 1933, R–12 was used often in centrifugal compressors for air conditioning applications.

The years to follow brought on the development of many more chlorofluorocarbon refrigerants. Listed below are these refrigerants with the dates of introduction to the commercial market along with other important dates.

| | |
|---|---|
| 1930 | Development of chlorofluorocarbons |
| 1931 | R-12 |
| 1932 | R-11 |
| 1933 | R-114 |
| 1934 | R-113 |
| 1936 | R-22 |
| 1961 | R-502   An azeotropic mixture of HCFC-22 and CFC-115 |
| 1974 | Ozone depletion theory. |
| 1978 | Ban on non-essential aerosols.<br>Global warming came into view. |
| 1985 | Stratospheric ozone hole discovered. |
| 1987 | Montreal Protocol.<br>Current tax rate schedule on CFC refrigerants. |
| 1990 | Clean Air Act Amendments.<br>Refrigerant production cuts and bans. |
| July 1992 | Unlawful to vent CFCs and HCFCs into atmosphere. |
| Nov. 15 1995 | Unlawful to vent alternative refrigerants (HFCs)<br>into atmosphere. |
| 1996 | Phaseout of CFC refrigerants. |
| 1996 | Freeze HCFC production. |
| 1997 | Kyoto protocol intended to reduce world wide<br>global warming gasses.<br>Global warming has become a major environmental issue. |
| 1998 | EPA "proposed" more strict regulations on recovery/<br>recycling standards equipment leak rates, and alternative<br>refrigerants. |
| 2020 | No production and no importing of HCFC-22 (R-22). |
| 2030 | No production and no importing of any HCFC refrigerant. |

# PHASE–OUT CHART

| Montreal Protocol | | United States | |
|---|---|---|---|
| Year by which Developed Countries Must Achieve % Reduction in Consumption | % Reduction in Consumption, Using the Cap as a Baseline | Year to be Implemented | Implementation of HCFC Phaseout through Clean Air Act Regulations |
| 2004 | 35.0% | 2003 | No production and no importing of HCFC–141b |
| 2010 | 65% | 2010 | No production and no importing of HCFC–142b and HCFC–22, except for use in equipment manufactured before 1/1/2010 (so no production or importing for NEW equipment that uses these refrigerants) |
| 2015 | 90% | 2015 | No production and no importing of any HCFCs, except for use as refrigerants in equipment manufactured before 1/1/2020 |
| 2020 | 99.5% | 2020 | No production and no importing of HCFC–142b and HCFC–22 |
| 2030 | 100% | 2030 | No production and no importing of any HCFCs |

**7**

# Appendix II

## Glossary

**Absolute pressure**

The force of a gas against a surface, measured in pounds per square inch. Equal to gauge pressure plus 14.7 (atmospheric pressure).

**Absolute temperature**

Temperature scale using absolute zero (–460°F) as 0°. Both the Kelvin and Rankin scales are absolute.

**Absolute zero**

The theoretical temperature at which molecular motion ceases; equal to –460°F (–275°C). The lowest possible temperature.

**Accumulator**

A device at the evaporator outlet that prevents liquid refrigerant from returning to the compressor.

**AEV**

Automatic expansion valve.

**Air conditioner**

A device that modifies the temperature, humidity, cleanliness or general quality of air.

**Air conditioning**

The science of controlling the temperature, humidity, cleanliness or general quality of air.

**Alkylbenzene**

An organic lubricant that's made from the raw chemicals propylene, a colorless hydrocarbon gas, and benzene, a colorless liquid hydrocarbon.

**Ambient temperature**

Temperature of the air around an object. *Ambient* comes from a Latin word that means "to surround."

**Antidote**

A substance that counteracts the effects of a poison.

**Aperture**

A service connection or port used to access a sealed refrigeration system, like a clamp–on piercing valve.

**Appliance**

A broad term used for electrical devices, including air–conditioning and refrigeration units (refrigerator, freezer, central air conditioner, walk–in cooler, or centrifugal chiller).

**Atmosphere**

Air; the gases surrounding the earth.

**Atmospheric pressure**

The pressure caused by the weight of the air above a certain point. Normal atmospheric pressure at sea level is about 14.7 pounds per square inch.

**Atom**

The smallest unit of an element. Every atom is made up of a positively charged *nucleus* and a set of negatively charged *electrons* that revolve around the nucleus. The nucleus is made up of positively charged *protons* and *neutrons* that have no charge. Atoms link together to form *molecules.*

**Atomize**

To reduce to a fine mist, or minute particles.

**Automatic expansion valve (AEV or AXV)**

A refrigerant metering device that maintains a constant evaporator inlet pressure.

**Azeotrope**

A constant–boiling mixture. A mixture of two liquids that boils at constant composition—the vapor's composition is the same as the liquid's. When the mixture boils, at first the vapor has a higher proportion of one component than is present in the liquid, so this proportion in the liquid falls over time. Eventually, maximum and minimum points are reached, at which the two liquids distill together with no change in composition. An azeotrope's composition depends on pressure. (See zeotrope.)

**Binary**

Anything made up of two parts. From a Latin word meaning "two by two."

**Boil**

To change from a liquid to a vapor.

**Boiling point**

The boiling temperature of a liquid.

**Boiling temperature**

The temperature at which a fluid changes from a liquid to a gas.

**British thermal unit (Btu)**

The amount of heat required to raise or lower the temperature of one pound of water one degree Fahrenheit.

**Btuh**

Btu per hour.

**Bubble point**

The liquid temperature at a given pressure of a refrigerant that has a discernable temperature glide.

**Bulb, sensing**

A fluid–filled bulb that responds to a temperature remote from its control.

**Calibrate**

To adjust the graduations of a measuring instrument.

**Capillary tube**

Small diameter (looped) tubing used as a metering device in refrigeration systems. Also called a cap tube.

**Change of state**

The transition from one of the three states of matter (gas, liquid or solid) to another.

**Charge**

1. refrigerant contained in a sealed system or in a sensing bulb, such as that of a thermostatic expansion valve. 2. to add refrigerant to a system.

**Chlorine: (CL)**

A chemical element used in the manufacturing of CFC and HCFC refrigerants.

**Chlorofluorocarbon (CFC)**

Any of several compounds made up of chlorine, fluorine and carbon. CFCs were used as aerosol propellants and refrigerants until they were found to be harmful to the earth's protective ozone layer.

**Commercial Refrigeration**

Refrigeration equipment utilized in the retail food and cold storage warehouse sectors.

**Compatible**

Capable of orderly, efficient integration and operation with other elements in a system.

**Compound**

In chemistry, a substance that contains two or more elements in definite proportions. Only one molecule is present in a compound.

**Compound gauge (Low–side gauge)**

A device that senses and measures pressures above and below atmospheric pressure (0 psig).

**Compression**

The squeezing of a gas to reduce its volume.

**Compression gauge (High–side gauge)**

A device used to measure pressures above atmospheric pressure, in pounds per square inch.

**Compression ratio**

The ratio of absolute compressor discharge pressure to absolute suction pressure.

**Compressor**

A mechanical pump in a refrigeration system that intakes refrigerant vapor and raises its temperature and pressure to the point where it can be condensed for re–use.

**Condense**

To change from a vapor to a liquid. From a Latin word that means "to thicken."

**Condenser**

A heat exchanger in which compressed refrigerant vapor is cooled until it becomes a liquid.

**Condensing pressure**

The pressure at which a vapor liquefies.

**Condensing temperature**

The temperature at which a gas becomes a liquid; varies with pressure.

**Configuration**

An arrangement of elements or parts in a system.

**Contaminants**

Dirt, moisture, or any other substance that is foreign to a refrigerant.

**Critical Temperature**

The highest temperature a gas can have and still be condensable by pressure.

**DeMinimus**

Minimum. The smallest quantity, number, or degree possible or permissible.

**Desuperheater**

An accessory that lowers the temperature of superheated vapor to saturation or near saturation conditions.

**Dew point**

The vapor temperature at a given pressure of a refrigerant that has a discernable temperature glide.

**Disposable cylinder**

A one–trip refrigerant cylinder; not to be refilled.

**Discharge line**

The line between the compressor discharge port and condenser. Also called the hot gas line.

**Discharge pressure**

Refrigerant pressure at the compressor outlet.

**Equalizer valve**

A device that regulates the flow of gases or liquids. It's used to balance pressures on either side of some recovering machines.

**Ester Oil**

An oil used with hydrofluorocarbon (HFC) refrigerants.

**Evacuate**

To remove air (gas) and moisture from a refrigeration or air conditioning system.

**Evaporating temperature**

The temperature at which a liquid will vaporize at a given pressure.

**Evaporation**

The conversion of a liquid to a vapor or gas.

**Evaporator**

A tubing coil in which a volatile liquid vaporizes, absorbing heat.

**Expansion valve**

A metering device used in refrigeration and air–conditioning applications that separates the system's low and high sides.

**Filter–Drier**

A device designed to remove moisture, acid and other impurities from the refrigerant.

**Fluorine**

A gaseous or liquid chemical element. It is a member of the halide family. Abbreviated: F.

**Fluorocarbon**

A molecule that contains fluorine and carbon atoms.

**Fractionation**

When one or more refrigerants of the same blend leak at a faster rate than other refrigerants in the blend, changing the composition of the blend. *Fractionation is possible only when liquid and vapor exist at the same time.*

**Freon**

The trademark for a family of fluorocarbon refrigerants manufactured by the DuPont Company.

**Fully halogenated CFC**

When all the hydrogen atoms in a hydrocarbon molecule are replaced with chlorine or fluorine atoms.

**Gauge manifold**

A tool on which gauges, valves and lines are installed to sense pressures in various parts of a refrigeration system.

**Gauge pressure**

A fluid pressure scale in which atmospheric pressure equals zero pounds and a perfect vacuum equals 30" mercury.

**Gauge, vacuum**

A gauge that measures pressures below atmospheric pressure.

**Global warming**

Often called the *greenhouse effect*. In global warming, tropospheric pollutants like CFCs, HCFCs, HFCs, carbon dioxide, and carbon monoxide, absorb and reflect the earth's infrared radiation. This causes re–radiation back to the earth, and a gradual increase in the earth's average temperature.

**Halogen**

Any of the five chemically related nonmetallic elements that include fluorine, chlorine, bromine, iodine, and astatine.

**Halogenate**

To cause some other element to combine with a halogen.

**Head pressure**

High–side pressure in a refrigeration system; pressure from the compressor discharge to the metering device.

**Heat**

The form of energy associated with molecular vibration.

**Heat transfer**

The movement of thermal energy via conduction, convection or radiation.

**Hermetic**

Totally sealed, especially against the escape or entry of air. In HVACR applications, it means sealed by gaskets or welds, as in refrigeration compressors.

**Hermetically sealed**

Any object or substance confined in an air tight container. A refrigeration system is hermetically sealed.

**High–pressure appliance**

An appliance that uses a refrigerant with a boiling point between 50 degrees C and 10 degrees C at atmospheric pressure.

**High side**

Any part of a refrigeration system under high pressure; that section of a refrigeration system starting at the compressor discharge and extending to the metering device.

**HVAC**

Heating, ventilation, air conditioning.

**Hydrocarbon**

A molecule that contains hydrogen and carbon atoms. An organic compound containing only hydrogen and carbon.

**Hydrochlorofluorocarbons (HCFCs)**

Molecules created when some of the hydrogen atoms in a hydrocarbon molecule are replaced with chlorine or fluorine atoms. Because they have a shorter life than CFCs, HCFCs are less harmful the CFCs to stratospheric ozone.

**Hydrofluorocarbons (HFCs)**

Molecules created when some of the hydrogen atoms in a hydrocarbon are replaced with fluorine. Because HFCs contain no chlorine, they don't destroy ozone but the contribute to global warming.

**Hydrostatically tested**

A process used to test the bursting points of cylinders or tanks (pressure vessels). They're filled with fluid, tightly closed, then subjected to a calibrated pressure.

**Hygroscopic**

Readily absorbs and retains moisture, as from the atmosphere.

**Inch of mercury (Hg)**

Unit of measurement for pressures below zero psig (atmospheric); equal to approximately 0.5 psi.

**Incompatible**

Not suited to be used together; not in harmony or agreement.

**Inert**

An inert chemical is one that shows no chemical activity except under extreme conditions. For example nitrogen is relatively non–reactive.

**Isomers**

Molecules that have the same numbers of the same atoms, but the atoms are arranged differently in their structure. Even though isomers of the same compound have equal numbers of atoms of the same element, they have very different physical properties.

**Latent heat**

Heat that can't be measured with a thermometer.
Latent heat is generated when substances change states.

**Leak detector**

Any device or substance that locates fluid (especially refrigerant or gas) leaks by reacting in their presence.

**Liquid**

The state of matter that takes the shape of its container, except for the top surface which is horizontal; a fluid state.

**Liquid line**

The refrigerant tubing extending from the condenser outlet to the metering device.

**Low–Loss Fittings**

Any device that connects hoses, appliances, or recovery or recycling machines, and that is designed to close automatically or to be closed manually when disconnected.

**Low Pressure Appliance**

An appliance that uses a refrigerant with a boiling point above 50 degrees F at atmospheric pressure. Evaporative pressure is below atmospheric.

**Low side**

The low–pressure side of a refrigeration system, from the metering device outlet to the compressor suction valve.

**Low–side pressure**

Back pressure; the pressure of the suction side of a refrigeration system.

**Lubricant**

Any substance that reduces friction.

**Malignancy**

Abnormal mass of new tissue growth that serves no function in the body and that threatens life or health.

**Metering device**

A valve or small diameter tube that restricts fluid flow.

**Micron gauge**

An instrument that measures very high vacuums in thousandths of millimeters.

**Miscible**

Capable of being mixed in all proportions.

**Mixture**

A blend of two or more components that do not have a fixed proportion to each other and that, however well blended, keep their individual chemical characteristics. Unlike compounds, mixtures can be separated by physical methods like distillation. *Examples are azeotropic and near–azeotropic blended refrigerants.*

**Moisture indicator**

An instrument used to measure a refrigerant's moisture content.

**Molecule**

A stable configuration of atoms held together by electrostatic and electromagnetic forces. A molecule is the simplest structural unit that displays a compound's characteristic physical and chemical properties.

**Montreal Protocol**

An agreement signed in 1987 by the United States and 22 other countries, and updated several times since then, to control releases of ozone–depleting substances (ODS) such as CFCs and HCFCs, and eventually phase out their use.

**Near–Azeotropic blend**

A blend that acts very much like an azeotrope, but has a small volumetric composition change and temperature glide as it evaporates and condenses.

**Nitrogen**

A colorless, odorless, relatively inert gas used to pressure test and purge refrigerant piping.

**Nomenclature**

A system of special terms or symbols, like those used in science. The numbering system used to name different refrigerants. Nomenclature comes from the Latin word "nomenclator," a slave who accompanied his master to tell him the names of people he met.

**Noncondensable gas**

Gas that does not change to a liquid at operating temperatures and pressure and therefore can not be condensed.

**Nonmiscible**

When two substances, such as oil and water, are incapable of mixing.

**Oil**

1. a liquid lubricant. 2. a heavy, liquid fuel.

**Operating pressure**

The normal refrigerant pressure during a unit's on–cycle.

**Organic**

Something derived from living organisms.

**Orifice**

An opening or hole; an inlet or outlet.

**Oxidation**

Any chemical reaction where a substance gives up electrons–as when a substance combines with oxygen. ***Burning is an example of fast oxidation; rusting is an example of slow oxidation.***

**Oxidize**

A corrosive chemical reaction caused by exposure to oxygen gas; like rust (iron oxide) or copper oxide (which forms on or inside copper tubing).

**Ozone**

$O_3$; a form of oxygen created by electrical discharge in air; used to eliminate odors but toxic in concentration.

**Ozone Depletion**

Happens when ultraviolet "C" radiation in the stratosphere breaks CFC and HCFC refrigerants into their atomic elements—Chlorine, fluorine and hydrogen atoms. Chlorine atoms react with and destroy stratospheric ozone, which protects earth's human and other life forms from the sun's harmful ultraviolet "A & B" radiation.

**Partially halogenated HCFC**

When not every hydrogen atom in a hydrogen molecule is replaced with chlorine or fluorine atoms.

**Permeability**

An object's or substance's ability to be penetrated.

**Phosgene**

A poisonous gas that is formed when halide refrigerants are burned.

**Polyalkylene glycols (PAGs)**
A very hygroscopic refrigeration lubricant for use with HFC refrigerants. Used often in automotive air conditioning systems when employing HFC refrigerants. PAGs are incompatible with chlorine and have very high molecular weights.

**Polymer**
A long molecule made up of a chain of smaller, simpler molecules.

**Polyolesters**
Polyolesters have stable five–carbon neopentyl alcohols which when mixed with fatty acids, will form the polyol ester family. A popular synthetic lubricant for use with HFC refrigerants. Used as a jet engine lubricant for years.

**Pounds per square inch   (PSI)**
A unit of pressure equal to the pressure resulting from a force of 1 pound applied uniformly over an area of 1 square inch.

**Pounds per square inch, absolute (PSIA)**
Pressure measurement that starts with atmospheric pressure then adds the pressure being measured. Example: 14.7 + PSIG = PSIA

**Pounds per square inch, gauge   (PSIG)**
Pressure measured on a manometer (gauge).

**Pressure**
The force exerted by a fluid on its container, per unit area.

**Pressure drop**
The loss of fluid pressure due to friction created by valves and/or uneven inner surfaces in a piping system.

**Pressure, saturation**
At a given temperature, the pressure at which a liquid and its vapor or a solid and its vapor can coexist in stable equilibrium.

**Pressure–temperature relationship**
The constant, predictable relationship between the pressure and temperature of a given liquid and gas mixture under saturated conditions. (See Bubble and Dew Point)

**Pressure vessel**
A holding device that maintains a certain "force per unit of area." Example: Refrigerant cylinder.

**Process stub**
A tube that extends from the compressor or filter drier of a hermetic system. It is used to gain access to the sealed system.

**Proprietary**

Sole ownership of property, a business, an item of labor, or an object that extends legal ownership rights. From Latin words that mean "property" and "one's own."

**Pressurize**

To introduce refrigerant or inert gas into a system in order to check for leaks.

**Psi**

Pounds per square inch.

**Psia**

Pounds per square inch, absolute.

**Psig**

Pounds per square inch, gauge.

**Pump**

A machine used for creating fluid flow.

**Quick–disconnect fittings**

Fittings used on refrigerant hoses that seal automatically when removed from an appliance. *Quick–disconnect fittings will help reduce refrigerant losses when removing hoses.*

**Receiver**

A refrigeration system component installed in the liquid line. It is designed to make a space for liquid refrigerant flow due to the closing action of a self–regulating metering device.

**Reclaim**

The process of returning recovered refrigerant to new product specifications. This is usually performed at a reprocessing facility.

**Recovery**

The process of removing refrigerant from a system, as is, and placing it in a container.

**Recycling**

The process of filtering recovered refrigerant to reduce contaminant levels.

**Refrigerant**

Any substance that transfers heat from one placing to another, creating a cooling effect.

**Refrigerant control**

Any valve or device that regulates the flow of refrigerant through a system.

**Refrigerant migration**

The condensation of refrigerant vapor at the coldest point in the system during the off cycle, usually occurs in the compressor.

**Refrigeration**

The process of cooling by removing heat.

**Refrigeration cycle**

A process during which a refrigerant absorbs heat at a relatively low temperature and rejects heat at a higher temperature.

**Residue**

A substance left over at the end of a process. For example, *residual oil* is the low–grade oil product left after gasoline is distilled.

**Retrofit**

To furnish with new equipment or parts that weren't available when a device or system was first manufactured.

**Saturated pressure**

The force in a pressure vessel that matches the temperature of a certain contained gas at a condition where any removed heat would cause condensation, and added heat would cause evaporation.

**Saturation temperature**

The temperature at which a liquid turns to vapor or a vapor turns to a liquid.

**Schrader valve**

Valves that use a valve core, like a tire–valve stem, to gain access to a sealed system. *Schrader valves help HVACR technicians recover refrigerant.*

**Seasonal energy efficiency ratio (SEER)**

A measure of cooling capacity.

**Self–contained recovery system (Active)**

Has its own means to draw the refrigerant out of the system (built-in pump).

**Sensible heat**

Heat energy which, when added to or removed from a substance, causes a rise or fall in temperature.

**Short cycling**

Continual starting and stopping of a system over a shorter–than–normal time period, due to a malfunction.

**Sight glass**

A transparent port (in the liquid line) that permits internal observation of a closed system.

**Solenoid**

A coil of wire surrounding a movable iron core. When current is applied, the resulting magnetic field moves the core which, in turn, can operate a switch or valve.

**Solenoid valve**

A control device that is opened and closed by an electrically energized coil.

**Soluble**

A substance that can be dissolved in a given liquid.

**Split system**

An air conditioning system in which the evaporator and condensing unit are located separately.

**Stratosphere**

The atmosphere between 7 and 30 miles above the earth where a layer of ozone filters out harmful ultraviolet light.

**Subcooling**

A liquid below its saturation temperature for a certain saturation pressure.

**Suction line**

The refrigerant piping from the evaporator outlet to the compressor inlet.

**Suction pressure**

The refrigerant pressure in the suction line, at the compressor inlet.

**Suction side**:

The low–pressure portion of the refrigeration system, from the metering valve to the compressor inlet.

**Superheat**

A vapor above its saturation temperature for a certain saturation pressure.

**Superheated gas**

A gas that has been heated to a temperature above its boiling temperature.

**Synthetic**

Produced artificially. In chemistry, forming a compound from its parts.

**System–Dependent Recovery Equipment (Passive)**

Refrigerant recovery equipment that requires the assistance of components contained in an appliance to remove the refrigerant from the appliance.

**Temperature**

The intensity of heat energy as measured by a thermometer.

**Temperature difference**

The number of degrees between two temperatures; determines the speed of heat transfer from the hotter to the colder substance.

**Temperature Glide**

Range of condensing or evaporating temperatures for one pressure.

**Ternary**

Having three elements, parts, or divisions.

**Thermodynamics**

The science of heat and energy.

**Thermometer**

A device that measures heat intensity or temperature.

**Thermostat**

A temperature–sensitive electrical circuit control.

**Thermostatic expansion valve (TEV or TXV)**

A valve that controls the flow of refrigerant. It is operated by evaporator temperature and pressure.

**Throttle**

To reduce or cut off fluid flow.

**Throttling valve**

A valve designed to atomize liquid refrigerant.

**Ton (refrigeration)**

The amount of heat absorbed in melting one ton of ice in 24 hours. Equal to 288,000 Btu per day, 12,000 Btu per hour or 200 Btu per minute.

**Total charge**

The amount of refrigerant needed for a system's operation, plus the refrigerant needed to fill its lines, which can be of varying length.

**Total equivalent warning impact (TEWI)**

A unit of measurement that assesses the total effect CFCs, HCFCs, and HFCs have on global warming.

**Total heat**

The sum of the latent and sensible heat contained in a substance; see enthalpy.

**Toxic**

Poisonous; hazardous to life.

**Toxicology**
The study of poisons, their effects and antidotes.

**Transition**
The process of changing from one state or form to another.

**Troubleshooting**
The process of observing a malfunctioning system's operation and diagnosing the cause of the malfunction.

**Troposphere**
The lowest level of the atmosphere—from the ground to seven miles above the earth—where ultraviolet rays from the sun react with pollution and smog to form ozone.

**Tube**
A small–diameter, flexible pipe, usually of copper or aluminum.

**Ultraviolet radiation**
Radiation in the part of the electromagnetic spectrum where wavelengths are shorter than visible violet light but longer than X–rays. UV radiation causes cancer.

**Undercharged**
A refrigeration system that is short of refrigerant.

**Vacuum**
Any pressure below atmospheric pressure.

**Vacuum pump**
A vapor pump capable of creating the degree of vacuum necessary to evaporate moisture near room temperature.

**Vent**
An opening that permits the escape of unwanted gases.

**Vapor**
A gas that is usually near saturation conditions.

**Vapor charged**
Lines and components that are filled with refrigerant before shipment from the factory.

**Vapor pressure**
Pressure applied to a saturated liquid.

**Very High Pressure Appliance**
An appliance which uses a refrigerant with a boiling point below –50° C. at atmospheric pressure.

**Virgin refrigerant**
New, original, non–recycled or reclaimed refrigerant.

**Wiring diagram**
An illustration of the various connections and components in an electrical system.

**Zeotrope**
Refrigerant blends that change volumetric composition and saturation temperatures as they evaporate or condense at constant pressures. Zeotropes have a temperature glide as they evaporate and condense. (Zeotrope and non–azeotrope are synonyms)

**Zero (absolute pressure and temperature)**
In absolute measuring systems, the absence of a condition, e.g., heat or pressure. In non–absolute systems, an arbitrary starting point for measurement.